Textbook of Agricultural Heritage

The Author

Professor Manas Mohan Adhikary (b.1951), a retired Professor of Agricultural Extension, Faculty of Agriculture, Bidhan Chandra Krishi Viswavidyalaya, West Bengal was attached with the profession of teaching, research and extension for about thirty-seven years. His immense potentiality of academic activities and capability to work with the people, sense of research and extension works helped him to serve as Dean, Faculty of Agriculture, Registrar, Bidhan Chandra Krishi Viswavidyalaya, Vice-Chancellor, Bidhan Chandra Krishi Viswavidyalaya, and Director (In charge), Directorate of Extension Education in the same University as well as Head of the Department of Agricultural Extension. To his credit there are 98 numbers of scientific research papers published in National & International Journals of repute, a good score of chapters and 24 numbers of books in related areas.

Textbook of Agricultural Heritage

(As Per ICAR's 5th Deans' Committee Recommendation)

Prof. Manas Mohan Adhikary

2020

Daya Publishing House®

A Division of

Astral International Pvt. Ltd.

New Delhi – 110 002

ISBN: 9789390371921(Int. Edition)

Publisher's Note:

Every possible effort has been made to ensure that the information contained in this book is accurate at the time of going to press, and the publisher and author cannot accept responsibility for any errors or omissions, however caused. No responsibility for loss or damage occasioned to any person acting, or refraining from action, as a result of the material in this publication can be accepted by the editor, the publisher or the author. The Publisher is not associated with any product or vendor mentioned in the book. The contents of this work are intended to further general scientific research, understanding and discussion only. Readers should consult with a specialist where appropriate.

Every effort has been made to trace the owners of copyright material used in this book, if any. The author and the publisher will be grateful for any omission brought to their notice for acknowledgement in the future editions of the book.

Published by : **Daya Publishing House®**
A Division of
Astral International Pvt. Ltd.
– ISO 9001:2015 Certified Company –
4736/23, Ansari Road, Darya Ganj
New Delhi-110 002
Ph. 011-43549197, 23278134
E-mail: info@astralint.com
Website: www.astralint.com

Digitally Printed at : **Replika Press Pvt. Ltd.**

Preface

Human achievements during the Stone Age are both fascinating and fundamental. Our knowledge of the Stone Age is limited however archeologist have been creative in their interpretation of tool remains and other evidence, such as cave paintings and burial sites, that Stone Age people produced in various parts of the world. What people accomplished during this long period of prehistory remains essential to human life today. Our ability to make and manipulate tools depends directly on what our Stone Age ancestors learned about physical matter. However, it was the invention of Agriculture that moved the human species toward more elaborate social and cultural patterns that people today would recognize. With Agriculture human beings were able to settle in one place and focus on economic, political, and religious goals and activities along with increasing the number of people in the world.

Agriculture is quite possibly the most important advancement and discovery that humanity has made. It produces the one thing that we need the most: food. It has been around since 9500 BC, and can be the oldest sign of mankind's acumen and the development and evolving of our minds and creations. Agriculture has been mastered throughout hundreds of years and is one of our most important resources on Earth, along with water and fossil fuels. Although the older farming methods from ancient times seem somewhat mediocre and barbaric, they were very ingenious and advanced for that time period. Over thousands of years, we have improved the way agriculture is used, how land is cultivated, the various techniques of farming and irrigation, and the tools and mechanics used. Numerous things that we see as aboriginal today, such as using a hand plow, were extremely contemporary in ancient times, and played key roles in the development of man and society, since quick labor was not abundant before this time.

We are now extremely advanced in agriculture and irrigation and the tools used to farm and grow and harvest crops. We have learned from our past and ancestors how to grow and evolve in our methods and have advanced forward greatly. Agriculture has been around for about 11,000 years. Around 9.500 BC, the first signs of crops began to show up around the coastlines of the Mediterranean. Emmer and einkorn wheat were the first crops that started to show up in this area, with barley, peas, lentils, chick peas, and flax following shortly.

The present book has been designed to cover the undergraduate syllabus of Agricultural Heritage as per ICAR 5th Deans' Committee report. It covers the entire syllabus in one concise volume, which is spread over in 16 chapters *viz.* introduction to Indian agricultural heritage, ancient agricultural practices and present day agriculture, past and present status of agriculture and farmers, astronomy: prediction of monsoon, ancient soil classification and soil fertility, irrigation and water harvesting, description of Indian civilization and agriculture by travellers, Indian agriculture from past to modern era, plant protection through indigenous traditional method, crop voyage in India and world, history of rice, sugarcane and cotton, gardening in ancient and medieval period: arbori-horticulture orchard, vegetable farming: floriculture perfumes and medicinal plants, crop significance and its classification, importance, scope, researches and national agricultural setup, current scenario of Indian agriculture. Presentation of the book is simple, lucid and unambiguous for easy understanding of the subject matter. It is hoped that this book will be greater use for undergraduate students of agriculture, teachers, extension workers and all other interested in agricultural heritage. We wish to express our deep sense of gratitude to the pioneers and authors whose illustrations and reviewed work have been utilized freely in the preparations of this book. In spite of the best efforts, it is possible that some errors may have occurred into the compilation and editing of the book. Further queries, suggestions and criticisms for improvement of the book are always welcome and shall be thankfully acknowledged.

We are highly thankful to publishers Astral International Pvt. Ltd., New Delhi for his keen interest shown in preparation and publishing of this book so efficiently and promptly.

Prof. Manas Mohan Adhikary

Contents

Chapter 1

Introduction to Indian Agricultural Heritage

Introduction: An Overview

- Our agriculture has lot of inherited sustainable practices passed from one generation to other generation. And also agriculture in India is not an occupation; it is a way of life for many Indian populations. Hence the present day generation should be aware about our ancient and traditional agricultural systems and practices. History denotes the continuous record of past events, where as heritage indicates the inherited values carried from one generation to other generation. Agricultural heritage denotes the values and traditional practices adopted in ancient India, which are more relevant for present day system.
- Indian agriculture began by 9000 BCE as a result of early cultivation of plants, and domestication of crops and animals. Settled life soon followed with implements and techniques being developed for agriculture. Double monsoons led to two harvests being reaped in one year. Indian products soon reached the world via existing trading networks and foreign crops were introduced to India. Plants and animals–considered essential to

Globally Important Agricultural Heritage Systems (GIAHS), as defined by the FAO (Food and Agriculture Organization of the UNO), are: "Remarkable land use systems and landscapes which are rich in globally significant biological diversity evolving from the co-adaptation of a community with its environment and its needs and aspirations for sustainable development".

their survival by the Indians–came to be worshiped and venerated. The middle ages saw irrigation channels reach a new level of sophistication in India and Indian crops affecting the economies of other regions of the world under Islamic patronage. Land and water management systems were developed with an aim of providing uniform growth. Despite some stagnation during the later modern era the independent Republic of India was able to develop a comprehensive agricultural programme.

- Our heritage is unique than any other civilization. As a citizen of India, we must feel proud about our rich cultural heritage. Agriculture in India is not of recent origin, but has a long history dating back to Neolithic age of 7500-4000 B.C. It changed the life style of early man from nomadic hunter of wild berries and roots to cultivator of land. Agriculture is benefited from the wisdom and teachings of great saints. The wisdom gained and practices adopted have been passed down through generations. The traditional farmers have developed the nature friendly farming systems and practices such as mixed farming, mixed cropping, crop rotation *etc.* The great epics of ancient India convey the depth of knowledge possessed by the older generations of the farmers of India.

Need for Studying Agricultural Heritage

- Our agriculture has lot of inherited sustainable practices passed from one generation to other generation. And also agriculture in India is not an occupation; it is a way of life for many Indian populations. Hence, the present day generation should be aware about our ancient and traditional agricultural systems and practices. This will enable us to build the future research strategy also.
- India has made tremendous progress in agriculture and its allied fields, but the emphasis on intensive use of inputs without considering their adverse impact of long term basis has created several problems related to sustainability of agriculture. Irrational use of chemical fertilizers, insecticides and exploration of natural resources is threatening the agro eco systems. Soil is getting impoverished, water and air getting polluted and there is an increasing erosion of plant and animal genetic resources. Therefore, attention in now shifting to sustainable form of agriculture. The indigenous technical knowledge (ITK) provides insight into the sustainable agriculture, because these innovations have been carried on from one generation to another as a family technology.
- There are several examples of valuable traditional technologies in India but unfortunately these small local systems are dying out. It is imperative that we collect, document and analyze these technologies so that the scientific principle/basis behind them could be properly understood. Once this done, it will be easier for us to further refine and upgrade them by blending them with the modern scientific technology.

Importance of Agricultural Heritage

Values and traditional practices adopted in ancient India which are more relevant for present day system. History denotes the continuous record of past events, where as heritage indicates the inherited values carried from one generation to other generation. Agricultural heritage denotes the values and traditional practices adopted in ancient India, which are more relevant for present day system.

- India has made tremendous progress in agriculture and its allied fields, but the emphasis on intensive use of inputs without considering their adverse impact of long term basis has created several problems related to sustainability of agriculture. Irrational use of chemical fertilizers, insecticides and exploration of natural resources is threatening the agro-ecosystems.
- Soil is getting impoverished, water and air getting polluted and there is an increasing erosion of plant and animal genetic resources. Therefore, attention in now shifting to sustainable form of agriculture.
- The indigenous technical knowledge (ITK) provides insight into the sustainable agriculture, because these innovations have been carried on from one generation to another as a family technology.
- There are several examples of valuable traditional technologies in India but unfortunately these small local systems are dying out. It is imperative that we collect, document and analyze these technologies so that the scientific principle/basis behind them could be properly understood.

Available Documents on agriculture during ancient and medieval period are summarized as under:

- Rigveda (c.3700 BC)
- Atharvaveda (c. 2000 BC)
- Ramayana (c.2000 BC)
- Mahabharata (c.1400 BC)
- Krishi-Parashara (c.400 BC)
- Kautilya's Artha-sastra (c.300 BC)
- Amarsimha's Amarkosha (c.200 BC)
- Patanjali's Mahabhasya (c.200 BC)
- Sangam literature (Tamils) (200 BC-100 AD)
- Agnipurana (c.400 AD)
- Varahamihir's Brhat Samhita (c. 500 AD)
- Kashyapiyakrishisukti (c.800AD)
- Surapala's Vrikshayurveda (c.1000 AD)

- Lokopakaram by Chavundaraya (1025 AD)
- Someshwardeva's Manasollasa (1131 AD)
- Saranghara's Upavanavioda (c.1300 AD)
- Bhavaprakasha-Nighantu (c.1500 AD)
- Chakrapani Mistra'sViswavallbha (c.1580 AD)
- Dara Shikoh's Nuskha Dar Fanni-Falahat (c.1650 AD)
- Jati Jaichand's dairy (1658-1714 AD)
- Anonymous Rajasthani Manuscript (1877 AD)
- Watt's Dictionary of Economic Products of India (1889-1893 AD)

Agricultural Heritage in India

- The modern society has lost sight of the importance of the traditional knowledge which had been subjected to a process of refinement through generations of experience. The ecological considerations shown by the traditional farmers in their farming activities are now-a-days is reflected in the resurgence of organic agriculture.
- The available ancient literature includes the four Vedas (rig, yajur, sama, atharvana), nineteen Brahmanas (A total of 19 Brahmanas are extant at least in their entirety: two associated with the Rigveda, six with the Yajurveda, ten with the Samaveda and one with the Atharvaveda.), Aranyakas, Sutra literature, Susruta Samhita, Charaka Samhita, Upanishads, the epices Ramayana and Mahabharata, Puranas (20), Buddhist and Jain literature, and texts such as Krishi-Parashara, Kautilya's Arthasastra, Panini's Ashtadhyayi, Sangam literature of Tamils, Manusmirti (laws), Varahamihira's Brihat Samhita (maths and astrology), Amarkosha, Kashyapiya-Krishisukti and Surapala's Vriskshayurveda. This literature was most likely to have been composed between 6000 BC to 1000 AD.
- The information related to the biodiversity and agriculture (including animal husbandry) are available in these texts. Rigveda is the most ancient literary work of India. It believed that Gods were the foremost among agriculturists. According to Amarakosha (a thesaurus of Sanskrit written by the Jain or Buddhist scholar Amarasimha), Aryans were agriculturists. Manu and Kautilya prescribed agriculture, cattle rearing and commerce as essential subjects, which the king must learn.
- According to Patanjali (compiler of the Yoga Sûtras) the economy of the country depended on agriculture and cattle-breeding. Plenty of information is available in 'Puranas', which reveals that ancient Indians had intimate knowledge on all agricultural operations. Some of the well known ancient classics of India are namely, Kautilya's Arthashastra';

Panini's 'Astadhyayi'; Patanjali's 'Mahabhasya'; Varahamihira's 'Brahat Samhita'; Amarsimha's 'Amarkosha' and Encyclopaedic works of Manasollasa. These classics testify the knowledge and wisdom of the people of ancient period. Technical books dealing exclusively with agriculture were Sage Parashara's 'Krishiparashara' in 1000 A.D.

- Other important texts are Agnipurana and Krishi Sukti attributed to Kashyap (500 A.D.). Ancient Tamil and Kannada works contain lot of useful information on agriculture in ancient India. Agriculture in India made tremendous progress in the rearing of sheep and goats, cows and buffaloes, trees and shrubs, spices and condiments, food and non-food crops, fruits and vegetables and developed nature friendly farming practices. These practices had social and religious undertones and became the way of life for the people. Domestic rites and festivals often synchronised with the four main agricultural operations of ploughing, sowing, reaping and harvesting.
- In the Rigveda, there is reference to hundreds and thousands of cows; to horses yoked to chariots; to race courses where chariot races were held; to camels yoked to the chariots; to sheep and goats offered as sacrificial victims, and to the use of wool for clothing. The famous Cow Sukta indicates that the cow had already become the very basis of rural economy. In another Sukta, she is defined as the mother of the Vasus, the Rudras and the Adityas, as also the pivot of Immortality. The Vedic Aryans appear to have large forests at their disposal for securing timber, and plants and herbs for medicinal purposes appear to have been reared by the physicians of the age, as appears in the Atharva Veda.
- The farmers' vocation was held in high regard, though agriculture solely depended upon the favours of Parjanya, the god of rain. His thunders are described as food-bringing. The four Vedas mention more than 75 plant species, Satapatha Bhrahamna mentions over 25 species, and Charkaa Samhita (300 BC) an Aayurvedic (Indian medicine) treatise-mentions more than 320 plants. Susruta (400 BC) records over 750 medicinal plant species. The oldest book, Rigveda (4000 BC) mentions a large number of poisonous and non poisonous aquatic and terrestrial, and domestic and wild creatures and animals. Puranas mention about 500 species of plants.
- The science of arbori-horticulture had developed well and has been documented in Surapala's Vrikshayurveda. Forests were very important in ancient times. From the age of Vedas, protection of forests was emphasized for ecological balance. Kautilya in his Artha Sastra (321-296 BC) mentions that superintendent of forests had to collect forest produce through the forest guards. He provides a long list of trees, varieties, of bamboos, creepers, fibrous plants, drugs and poisons, skins of various animals, *etc.*, that come under the purview of this officer. The preservation

of wild animals was encouraged and hunting as a sport was regarded as detrimental to proper development of the character and personality of the ruler, according to Manu (Manusmriti, 2nd Century BC). Specifically, in the Puranas (300-750 AD) the names of Shalihotra on horses and Palakapya on elephants have been found as experts in animal husbandry.

- ☆ Agricultural heritage systems can still be found throughout the world covering about 5 million hectares, which provide a vital combination of social, cultural, ecological and economical services to humankind. These "Globally Important Agricultural Heritage Systems-GIAHS" have resulted not only in outstanding landscapes of aesthetic beauty, maintenance of globally significant agricultural biodiversity, resilient ecosystems and a valuable cultural heritage.
- ☆ All these systems sustainably provide multiple goods and services, food and livelihood security for millions of poor and small farmers. The existence of numerous GIAHS around the world testifies to the inventiveness and ingenuity of people in their use and management of the finite resources, biodiversity and ecosystem dynamics, and ingenious use of physical attributes of the landscape, codified in traditional but evolving knowledge, practices and technologies. Whether recognized or not by the scientific community, these ancestral agricultural systems constitute the foundation for contemporary and future agricultural innovations and technologies. Their cultural, ecological and agricultural diversity is still evident in many parts of the world, maintained as unique systems of agriculture. Through a remarkable process of co-evolution of Humankind and Nature, GIAHS have emerged over centuries of cultural and biological interactions and synergies, representing the accumulated experiences of rural peoples. Some Ancient and Traditional Tools are given in figure 1.1a and 1.1b.

HISTORY OF FISHERIES

An Introduction

- ☆ Fisheries sector play an important role in the socio-economic development of farmers in the country. The sector has been recognized as a powerful income and employment generator as it stimulates growth of a number of subsidiary industries, and is a source of cheap and nutritious food besides being a foreign exchange earner.
- ☆ Most importantly, it is the source of livelihood for a large section of economically backward population of the country. The main challenges facing fisheries development in the country includes accurate data on assessment of fishery resources and their potential in terms of fish production, development of sustainable technologies for fin and shell fish

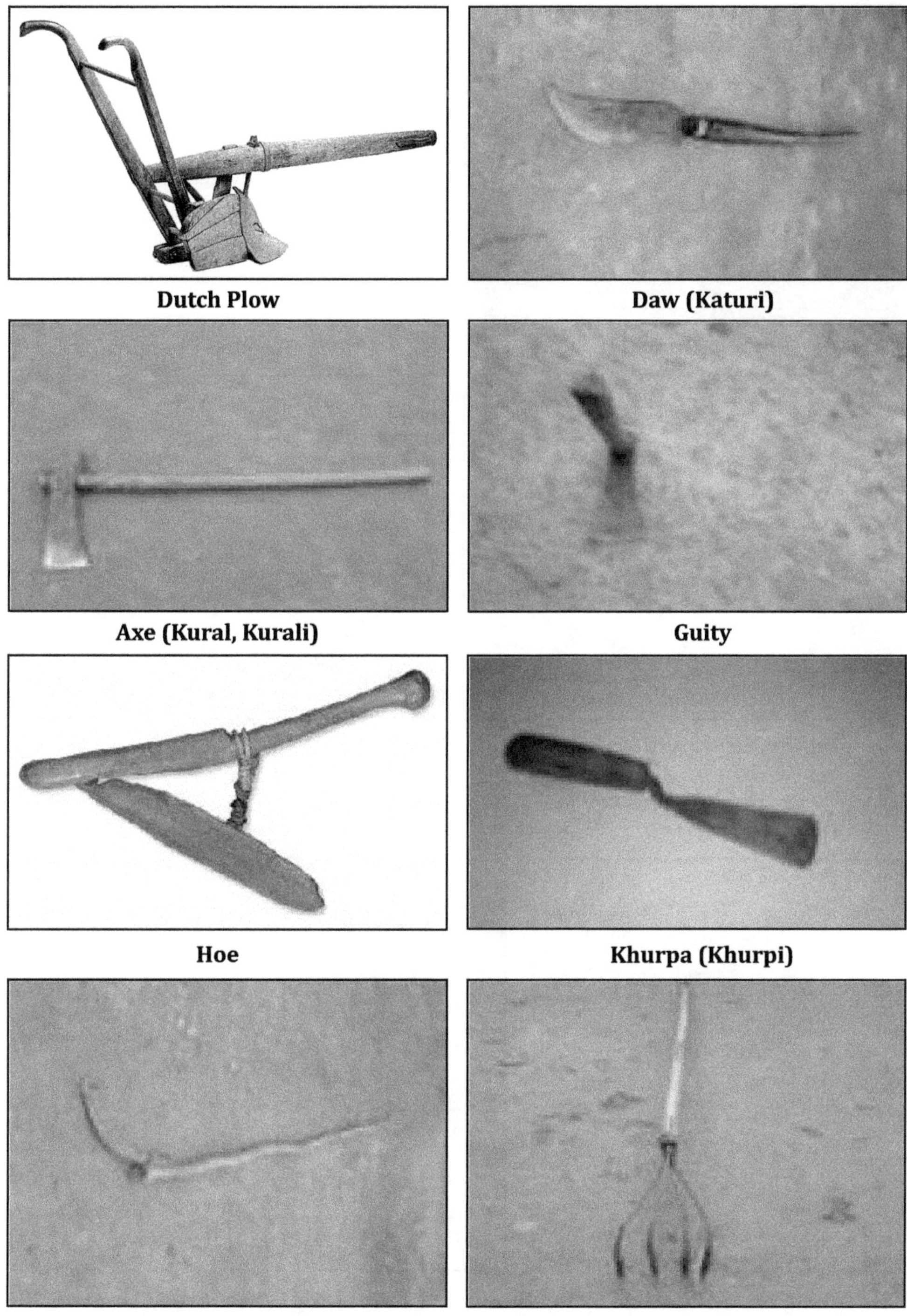

Dutch Plow **Daw (Katuri)**

Axe (Kural, Kurali) **Guity**

Hoe **Khurpa (Khurpi)**

Long Handle Dauli **Long Handle Weeder (Rice Khera)**

Figure 1.1a: Some Ancient and Traditional Agricultural Tools.

Plough (Dunckle)

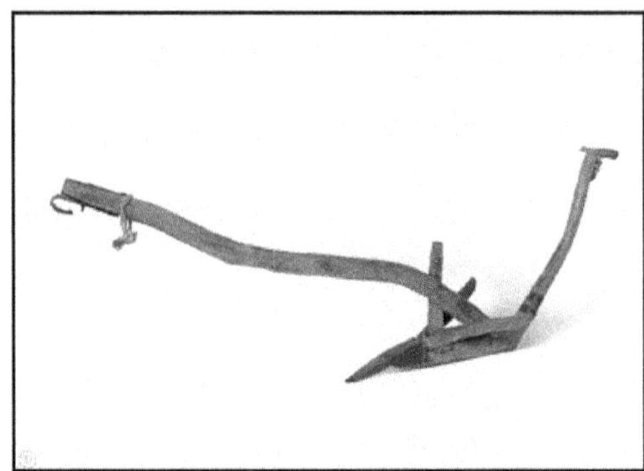

Wooden Plough

Scythe, Sickle and Mattock

Hoars

Threshing Paddy Pinnacle by Beating

Harvesting Wheat Crop using a Sickle

Figure 1.1b: Some Ancient and Traditional Agricultural Tools.

culture, yield optimization, harvest and post-harvest operations, landing and berthing facilities for fishing vessels and welfare of fishermen.

- There has been significant growth in fish production during recent years. India is now the third largest producer of fish and second largest producer of fresh water fish in the world.
- A network of 429 Fish Farmers Development Agencies (FFDA's) has been set up covering all the potential districts in all the States and Union Territories for propagating freshwater aquaculture.
- With a view to provide technical, financial and extension support to shrimp farmers in the small scale sector, 39 Brackish water Fish Farmers Development Agencies (FFDA's) have been set up in all the coastal States and the UT of Andaman and Nicobar Islands.

Fish as Food

- Malnutrition and starvation are the two serious problems being faced by millions of rural poor in most of the developing countries. The problem of malnutrition is in fact more serious and of a bigger dimension than the starvation problem and is caused mainly due to animal protein-deficient diets.
- Animal protein is essential for proper growth, repair and maintenance of body organs and tissues. Fish contain about 16-20 per cent protein compared to about 12 per cent in egg, 3.5 per cent in milk and 6-8 per cent in rice and wheat. Moreover, it is wholesome, tasty, highly nutritive and an excellent source of essential minerals, vitamins and essential amino acids.
- At present about 31 per cent of the total animal protein supply in the Asian region is in the form of fish protein. For the poorest segments of the population, fish is not only the most important animal protein source, but often the only one.

High Multiplication Capacity and Minimal Water Requirement

- The reproductive potential of fish compared to any other farmed animal is also very high. A kilogram of female cultivable carp species yields on an average about 0.1 million eggs, each of which has the potential to become 1 kg fish in about a year. No livestock animal possesses this magnitude of fecundity. Although fish needs water as a medium to survive and grow, it consumes minimal quantity of water compared with any livestock or agricultural crop. Fish also enriches the water with its voided metabolites thus making the water more productive for agriculture.

Low Energy Requirement for Protein Production

- Fish culture systems require a relatively less amount of energy for protein production than any other farming system. Carp culture, depending upon culture practices, requires energy at the rate of 22-468 KJ/g of protein production while a land animal farming system needs over 550 to 3 400 KJ/g.

Warm Water favours Fish Growth

- ☆ Fish are cold blooded or poikilothermic animals. In other words, they cannot maintain a constant and high body temperature like other livestock animals. Instead, their body temperature fluctuates according to the surrounding temperature.
- ☆ In warmer climates, their metabolism accelerates and they grow faster, while in colder climates, the metabolic rate slows down, resulting in a reduced rate of growth. In this way they save energy by not spending it for maintaining a higher and constant body temperature.

History

- ☆ Fishing is the activity of catching fishes. It has a history of over 35,000 years and may be an individual necessity or a collective undertaking involving large groups of men. Since the 16th century fishing vessels have been able to cross oceans in pursuit of fish and since the 19th century it has been possible to use large vessels and in some cases of fish on board.
- ☆ The term fishing may be applied to catching other aquatic animals such as Shellfish, Cephalopods, Crustaceans and Echinoderms. The term is not usually applied to catching aquatic mammals such as Whales.

Prehistory

- ☆ Fishing is an ancient practice that dates back at least to the Upper Paleolithic period which began about 40,000 years ago. Archaeological features such as shell middens, discarded fish bones and cave painting show that sea foods were important for survival and consumed in significant quantities.
- ☆ Spear fishing with barbed poles (harpoons) was widespread in Paleolithic times. Cosquer cave in Southern France contains cave art over 16,000 years old including drawings of seals which appear to have been harpooned. In the Old Stone Age (40,000B.C.) heaps of refuse of shellfish and sea fish were found at the dwelling sites of man near rivers and lakes.
- ☆ New Stone Age (10,000B.C.) has provided evidence of Salmon Smoking practices. Salting of Fish was probably started in the Bronze Age (3500).

Ancient History

- ☆ The Egyptians invented various implements and methods for fishing and these are clearly illustrated in tomb scenes, drawings, and papyrus documents. Fishing scenes are represented in Ancient Greek Culture depcicting the low social status of fishing. There is a wine cue dating from 500B.C. that shows a boy crouched on a rock with a fishing rod in his right hand and basket in his left.

- ☆ The Roman Marcus Terentius, arro (116-27B.C.) wrote in his book "De Re Rustica" about two kinds of ponds, freshwater ponds (dulces) owned by peasants for food and profit and salt water ponds (maritime or sales) owned by wealthy aristocrats who used them to entertain their guests.
- ☆ Cassidorus (AD.490-585) mentioned that live carps were taken from Danube to the Goth king at Theodoric at Ravenna in Italy.

16th Century Works

- ☆ Pierra Belon(1517-1575AD), described at least 110 fishes from the Eastern part of the Mediterranean in Europe in the publication entitled "De Aquatilibu slibri duo".
- ☆ H.Salviani (1514-1572) - 92 fishes of Italy (Rome) were described by him in "*Aquatilium animalium Historia*".
- ☆ G.Rondelet (1507-1557AD) About 197 marine and 47 freshwater fishes from Mediterranean find place in Rondelet's work "Libri De Piscibus Marinis" and "Universae aquatilium Historiae Pars altera".

17th Century Works

- ☆ After standard works on Ichthyology Belon, Rondelet *etc.*, W.Piso and G. Margrav(1611-1678AD.), described about 100 fishes from Brazil in the 4th vol. of "Historia naturalis Braziliae".

18th and 19th Century Works

- ☆ One of the epoch making Ichtyological works in 18th century was that of Peter Artidi(1705-1734), who has been aptly called as *Father of Ichthyology* owing to his valuable contribution towards laying down the firm foundation of the science of Ichthyology.
- ☆ After the untimely death of Peter Artidi, Carl Linnaeus, later Carolus Von Linnae(1707-1778), whom we know to-day as *Father of Taxonomy,* accomplished Aritidi's unfinished task in book form entitled "Artidi Ichthyologica". Linnaeus applied bionomial terms to the spp. properly described and classified by Artidi.
- ☆ Mark Eliezer Bloch (1723-1797), a German Physician, prepared a unique general system of fishes in which he arranged not only those described in his great works on the fishes of Germany but that which included Indian Fishes.
- ☆ The Genera of Fishes were published in four parts in the form of monographic series of Stanford University -The Genera of fishes Part 1, Part 2, Part 3, Part 4.

Indian History of Fisheries

- ☆ In India the interest in fish and fishery dates back to the Third Millennium BC, and evidences of fish being used as food are available from excavations of the Indus Valley Civilization.
- ☆ In 1127AD, the son of King Vikramaditya, King Somesvara composed a book recording the common sport fishes of India and grouped them into marine and riverine forms.
- ☆ As early as in 1822 Hamilton Buchman gave an excellent, illustrated taxonomic account of the 'Fishes of Gangetic System' and removed various confusions caused by regional names of fishes. After Hamilton, the epoch macking contribution of Francis Day's 'Fishes of India'(1878) and 'Fauna of British India, Burma and Ceylon'(1889).
- ☆ The credit for developing a sustained interest in highlighting the rich Indian Piscian funna largely goes to Rai Bahadur and Dr. Sundar Lal Hora, former Director of Zoological Survey of India. They gave to the study of taxonomy on ecological bias and while doing so a long series of papers have been published.
- ☆ J.S. Dutta Munshi and M.P. Srivastava of Bhagalpur University recently brought out an exhaustive treatise on 'Natural History Of Fishes And Systematics Of Freshwater Fishes of India(1988)'.
- ☆ Based on many researches on fishery science a brillent monograph on Fish Culture in India was brought out in 1957 by K.H. Alikunhi, the then scientist, Central Inland Fisheries Research Institute, Barrackpore (West Bengal).
- ☆ It was in 1975 when a comprehensive volume on Fish and fisheries of India was authored by Dr. V.G.Jhingaran, the then Director, CIFRI, Barrackpore(W.B.). This book has painstakingly blended into an integrated whole a great mass of scientific knowledge on Indian fishes and fisheries accumulated over decades.
- ☆ Dr. Jhingaran was honoured by **Rafi Ahmad kidwai** and India's most prestigious national award,**Padma Shri.**

History of Some Culture Practices

Composite Fish Culture

- ☆ The composite fish culture is a technology developed in India by the Indian Council of Agricultural Research in 1970's. In this system both local and important fish species, a combination of five or six fish spp., is used in a single fish pond. These species are selected so that they do not compete for food among them having different types of food habitats.
- ☆ Fish used in this system include Catla, and Silver Carp, which are surface feeders, Rohu a column feeder and Mrigal and Common Carp which are bottom feeders.

Cage Culture

- ☆ The cage culture originated about 200 years ago in Kampuchia (Combodia) from where it has spread to Indonesia, Thailand, India and other Asian Countries. This culture practice is quite peculiar in that the fish to be cultured are kept in cages of metal, mesh, or nylon mesh left in the flowing water.
- ☆ In the past few decades it has become a major source of aquaculture production, particularly highly esteemed, Salmon, Trouts, Yellow Tail, Sea Bass, Grouper spp. *etc.*

Pen Culture

- ☆ Pen culture was first started in **Indonesia.** The pen is considered as transitional structure between ponds and cages.
- ☆ The enclosures should be relatively small (2.0 to 7.0ha.). The areas with too much silt and decomposing organic matter should be avoided.

Monoculture

- ☆ This type of culture is aimed at to culture only one type of fish spp. In a well designed pond, tank, cage *etc.*

Mono Sex Culture

- ☆ In this case, only one member of the sex either male or female is cultured. The obvious advantage of such a practice is that all the energy of fish is utilized in growth.

History of Animal Husbandry

- ☆ Animal Husbandry is a branch of agriculture concerned with the domestication of, care for and breeding of animals such as dogs, cattle, horses, sheep, goats, pigs and other like creatures.

Birth of Animal Husbandry

- ☆ The first wild animal to be domesticated was the dog. Half-wild dogs, perhaps starting with young individuals, may have been tolerated as scavengers and killers of vermin, and being naturally pack hunters, were predisposed to become part of the human pack and join in the hunt.
- ☆ Prey animals, sheep, goats, pigs and cattle, were progressively domesticated early in the history of agriculture.
- ☆ Pigs were domesticated in Mesopotamia around 13,000 BC, and sheep followed, some time between 11,000 and 9,000 BC. Cattle were domesticated from the wild aurochs in the areas of modern Turkey and Pakistan around 8,500 BC.

- A cow was a great advantage to a villager as she produced more milk than her calf needed, and her strength could be put to use as a working animal, pulling a plough to increase production of crops, and drawing a sledge, and later a cart, to bring the produce home from the field. Draught animals were first used about 4,000 BC in the Middle East, increasing agricultural production immeasurably.
- In southern Asia, the elephant was domesticated by 6,000 BC. Horses occur naturally on the steppes of Central Asia, and their domestication, around 3,000 BC in the Black Sea and Caspian Sea region, was originally as a source of meat; use as pack animals and for riding followed. Around the same time, the wild ass was being tamed in Egypt. Camels were domesticated soon after this, with the Bactrian camel in Mongolia and the Arabian camel becoming beasts of burden. By 1000 BC, caravans of Arabian camels were linking India with Mesopotamia and the Mediterranean.

Ancient Civilisations

- In ancient Egypt, cattle were the most important livestock, and sheep, goats, and pigs were also kept; poultry including ducks, geese, and pigeons were captured in nets and bred on farms, where they were force-fed with dough to fatten them. The Nile provided a plentiful source of fish. Honey bees were domesticated from at least the Old Kingdom, providing both honey and wax.
- In ancient Rome, all the livestock known in ancient Egypt were available. In addition, rabbits were domesticated for food by the first century BC.

Medieval Husbandry

- The improvements of animal husbandry in the medieval period in Europe went hand in hand with other developments. Improvements to the plough allowed the soil to be tilled to a greater depth.
- Horses took over from oxen as the main providers of traction, new ideas on crop rotation were developed and the growing of crops for winter fodder gained ground. Peas, beans and vetches became common; they increased soil fertility through nitrogen fixation, allowing more livestock to be kept.

Columbian Exchange

- Exploration and colonisation of North and South America resulted in the introduction into Europe of such crops as maize, potatoes, sweet potatoes and manioc, while the principal Old World livestock – cattle, horses, sheep and goats – were introduced into the New World for the first time along with wheat, barley, rice and turnips.

Husbandry Systems

- Animal husbandry was part of the subsistence farmer's way of life, producing not only the food needed by the family but also the fuel, fertiliser, clothing, transport and draught power. Killing the animal for food was a secondary consideration, and wherever possible its products, such as wool, eggs, milk and blood (by the Maasai) were harvested while the animal was still alive.
- In the traditional system of transhumance, people and livestock moved seasonally between fixed summer and winter pastures; in montane regions the summer pasture was up in the mountains, the winter pasture in the valleys.
- Animals can be kept extensively or intensively. Extensive systems involve animals roaming at will, or under the supervision of a herdsman, often for their protection from predators. Ranching in the Western United States involves large herds of cattle grazing widely over public and private lands.
- Similar cattle stations are found in South America, Australia and other places with large areas of land and low rainfall. Similar ranching systems have been used for sheep, deer, ostrich, emu, llama and alpaca.
- In the uplands of the United Kingdom, sheep are turned out on the fells in spring and graze the abundant mountain grasses untended, being brought to lower altitudes late in the year, with supplementary feeding being provided in winter.
- In rural locations, pigs and poultry can obtain much of their nutrition from scavenging, and in African communities, hens may live for months without being fed, and still produce one or two eggs a week.

Feeding

- Animals used as livestock are predominantly herbivorous, the main exception being the pig which is an omnivore. The herbivores can be divided into "concentrate selectors" which selectively feed on seeds, fruits and highly nutritious young foliage, "grazers" which mainly feed on grass, and "intermediate feeders" which choose their diet from the whole range of available plant material. Cattle, sheep, goats, deer and antelopes are ruminants; they digest food in two steps, chewing and swallowing in the normal way, and then regurgitating the semidigested cud to chew it again and thus extract the maximum possible food value. The dietary needs of these animals is mostly met by eating grass. Grasses grow from the base of the leaf-blade, enabling it to thrive even when heavily grazed or cut.
- In many climates grass growth is seasonal, for example in the temperate summer or tropical rainy season, so some areas of the crop are set aside to be cut and preserved, either as hay (dried grass), or as silage. Other forage crops are also grown and many of these, as well as crop residues,

can be ensiled to fill the gap in the nutritional needs of livestock in the lean season.

- Extensively reared animals may subsist entirely on forage, but more intensively kept livestock will require energy and protein-rich foods in addition. Energy is mainly derived from cereals and cereal by-products, fats and oils and sugar-rich foods, while protein may come from fish or meat meal, milk products, legumes and other plant foods, often the by-products of vegetable oil extraction.
- Pigs and poultry are non-ruminants and unable to digest the cellulose in grass and other forages, so they are fed entirely on cereals and other high-energy foodstuffs. The ingredients for the animals' rations can be grown on the farm or can be bought, in the form of pelleted or cubed, compound foodstuffs specially formulated for the different classes of livestock, their growth stages and their specific nutritional requirements.

Breeding

- The breeding of farm animals seldom occurs spontaneously but is managed by farmers with a view to encouraging certain traits that are seen as desirable. These include hardiness, prolificness, mothering abilities, fast growth rates, low feed consumption per unit of growth, better body proportions, higher yields, better fibre qualities and other characteristics.
- Selective breeding has been responsible for some large increases in productivity. In 2007, a typical broiler chicken at eight weeks old was 4.8 times as heavy as a bird of similar age in 1957. In the thirty years to 2007, the average milk yield of a dairy cow in the United States nearly doubled.
- Artificial insemination and embryo transfer are frequently used today, not only as methods to guarantee that females breed regularly but also to help improve herd genetics. This may be done by transplanting embryos from high-quality females into lower-quality surrogate mothers – freeing up the higher-quality mother to be reimpregnated.

Animal Health

- Good husbandry, proper feeding, and hygiene are the main contributors to animal health on the farm, bringing economic benefits through maximised production. When, despite these precautions, animals still become sick, they are treated with veterinary medicines, by the farmer and the veterinarian.
- Animals are susceptible to a number of diseases and conditions that may affect their health. Some, like classical swine fever and scrapie are specific to one type of stock, while others, like foot-and-mouth disease affect all cloven-hoofed animals.

- Vaccines are available against certain diseases, and antibiotics are widely used where appropriate. At one time, antibiotics were routinely added to certain compound foodstuffs to promote growth, but this practice is now frowned on in many countries because of the risk that it may lead to antibiotic resistance.
- Governments are particularly concerned with zoonoses, diseases that humans may acquire from animals. Wild animal populations may harbour diseases that can affect domestic animals which may acquire them as a result of insufficient biosecurity. An outbreak of Nipah virus in Malaysia in 1999 was traced back to pigs becoming ill after contact with fruit-eating flying foxes, their faeces and urine.
- The pigs in turn passed the infection to humans. Avian flu H5N1 is present in wild bird populations and can be carried large distances by migrating birds. This virus is easily transmissible to domestic poultry, and to humans living in close proximity with them. Other infectious diseases affecting wild animals, farm animals and humans include rabies, leptospirosis, brucellosis, tuberculosis and trichinosis.

Chapter 2

Ancient Agricultural Practices and Present Day Agriculture

Ancient Agricultural Practices: An Introduction

- Agriculture began independently in different parts of the globe, and included a diverse range of taxa. At least eleven separate regions of the Old and New World were involved as independent centers of origin.
- Wild grains were collected and eaten from at least 20,000 BC. From around 9500 BC, the eight Neolithic founder crops–emmer wheat, einkorn wheat, hulled barley, peas, lentils, bitter vetch, chick peas, and flax–were cultivated in the Levant. Rye may have been cultivated earlier but this remains controversial. Rice was domesticated in China by 6200 BC with earliest known cultivation from 5700 BC, followed by mung, soy and azuki beans. Pigs were domesticated in Mesopotamia around 11,000 BC, followed by sheep between 11,000 BC and 9000 BC. Cattle were domesticated from the wild aurochs in the areas of modern Turkey and Pakistan around 8500 BC. Sugarcane and some root vegetables were domesticated in New Guinea around 7000 BC. Sorghum was domesticated in the Sahel region of Africa by 5000 BC. In the Andes of South America, the potato was domesticated between 8000 BC and 5000 BC, along with beans, coca, llamas, alpacas, and guinea pigs. Bananas were cultivated and hybridized in the same period in Papua New Guinea. In Mesoamerica, wild teosinte was domesticated to maize by 4000 BC. Cotton was domesticated in Peru by 3600 BC. Camels were domesticated late, perhaps around 3000 BC.

- The Bronze Age, from c. 3300 BC, witnessed the intensification of agriculture in civilizations such as Mesopotamian Sumer, ancient Egypt, the Indus Valley Civilisation of South Asia, ancient China, and ancient Greece. During the Iron Age and era of classical antiquity, the expansion of ancient Rome, both the Republic and then the Empire, throughout the ancient Mediterranean and Western Europe built upon existing systems of agriculture while also establishing the manorial system that became a bedrock of medieval agriculture. In the Middle Ages, both in the Islamic world and in Europe, agriculture was transformed with improved techniques and the diffusion of crop plants, including the introduction of sugar, rice, cotton and fruit trees such as the orange to Europe by way of Al-Andalus. After the voyages of Christopher Columbus in 1492, the Columbian exchange brought New World crops such as maize, potatoes, sweet potatoes, and manioc to Europe, and Old World crops such as wheat, barley, rice, and turnips, and livestock including horses, cattle, sheep, and goats to the Americas.
- Irrigation, crop rotation, and fertilizers were introduced soon after the Neolithic Revolution and developed much further in the past 200 years, starting with the British Agricultural Revolution. Since 1900, agriculture in the developed nations, and to a lesser extent in the developing world, has seen large rises in productivity as human labour has been replaced by mechanization, and assisted by synthetic fertilizers, pesticides, and selective breeding. The Haber-Bosch process allowed the synthesis of ammonium nitrate fertilizer on an industrial scale, greatly increasing crop yields. Modern agriculture has raised social, political, and environmental issues including water pollution, biofuels, genetically modified organisms, tariffs and farm subsidies. In response, organic farming developed in the twentieth century as an alternative to the use of synthetic pesticides.

Relevance of Heritage to Present Day Agriculture

Traditional Agriculture

- Considered 'backward' by the proponents of modern agriculture
- Dr John Voelcker, studied Indian agriculture practices and found them scientific
- It uses the irrigation system through wells
- Scientific rotation system is adopted (slow and quick growing crops, deep rooted and shallow plants, may co- exist)
- The ploughing and tilling retains the moisture of soil
- Mixing of soil with clay is done to grow other crops
- Weeding done by hand

- Traditional farms are small and farmers depend upon their own labor
- Environment friendly

Modern Farming

- Reduces soil fertility
- Artificial fertilizers used
- Deep ploughing by tractors results in soil erosion and loss of porosity
- Extensive use of pesticides
- Less biodiversity as farms are monoculture, growing the same crop and crop variety
- Exotic and hybrid varieties are grown and indigenous plant existence is threatened
- The food is contaminated with the chemicals used to produce it
- The supply and trading in agricultural inputs and produce is in the hands of large players which threatens the food security and reduces the leverage and importance of the farmer and the consumer
- The habitat of the wild plants and wild animals is being destroyed

Types of Agricultural Farming

- Shifting cultivation-The plot of land is cultivated temporarily and abandoned when it loses fertility.
- Subsistence farming-The farmer grows only to feed his own family.
- Intensive farming-Characterized by high input of capital, labour, fertilizers and pesticides relative to the land area. Increases crop production but also damages the environment.
- Extensive farming- Low input of materials and labour to preserve the ecological balance so that the land can be farmed indefinitely.
- Commercial agriculture-Farming intended for sale and done on a large scale with mechanised equipment.
- Dryland farming-Agricultural technique for non- irrigated cultivation of land with little natural rainfall.
- Monoculture-Agricultural practice of growing one crop over a large area. The processes can be standardized for greater efficiency. Result in surplus production of crop and depressed prices.
- Crop rotation- practice of growing dissimilar crops to improve soil structure and fertility by alternatively growing deep rooted and shallow rooted plants.

Factors Affecting Agriculture

- ☆ Small and fragmented landholdings
- ☆ Dependence on the monsoon
- ☆ Lack of international competitiveness of its produce
- ☆ Inadequate availability of electricity, fertilizers, irrigation and pesticides
- ☆ Poor access of the farmers to good roads, market infrastructure, refrigerated transportation of goods
- ☆ Conversion of agricultural land for residential and other land use purposes.

Organic Farming

- ☆ Organic Cultivation is a type of farming that does not involve usage of chemicals like chemical fertilizers and pesticides.
- ☆ Major shift has been observed in the farming culture, due to which several farmers have begun practicing this traditional method of cultivation
- ☆ Organic cultivation is proven as the means to produce safe foodstuffs and preserve the environment.
- ☆ It uses organic fertilizers, which are carbon based, and increase the productivity of plants
- ☆ The farms retain their fertility
- ☆ Organic fertilizers are bio-degradable so do not cause environmental pollution
- ☆ Nutrients are added after soil testing
- ☆ Organic seeds are used
- ☆ Organic fungicides, pesticides used
- ☆ Organic herbicides and mulching used to control the weeds.

Agricultural Achievements

- ☆ Nearly 50 years ago, a panel of experts in the agriculture field met in Italy (called the club of Rome) and predicted massive food shortages and death due to hunger. In India their predictions were supported by the then data which indicated that after 1947 (since Independence) the agriculture output fell a way short of the demand for food. Indian resilience proved them wrong.
- ☆ To-day, India is the largest product of milk (128 million tonnes) second largest producer of rice (over 100 million metric tonnes), wheat over 90 million tonnes, sugar 25 million tonnes, cotton 34 million bales and fruits and vegetables, over 200 million metric tonnes. Besides, it is a significant producer of a variety of spices, plantation crops, poultry and fishery products.

- How did all this come about? The government support was there but not enough. Green Revolution, a process which improved the food production significantly and developed with the support of the Indian Scientists likes of the caliber of Dr. M. S. Swaminathan and Dr. Kurien played a very big role in achieving this near Self Sufficiency. Currently agriculture accounts for 15 per cent of India's GDP, employs about 55 per cent of the work force and contributes to around 10 per cent of exports.

Agriculture is the most important sector of Indian Economy. Indian agriculture sector accounts for 18 per cent of India's gross domestic product (GDP) and provides employment to 50 per cent of the countries workforce. India is the world's largest producer of pulses, rice, wheat, spices and spice products. India has many areas to choose for business such as dairy, meat, poultry, fisheries and food grains etc. India has emerged as the second largest producer of fruits and vegetables in the world.

- Unfortunately, once we reached a certain level of output our government had become complacent. Funding all activities of agriculture like the subsidies have become a political issue. It is myopic and hurting. Our last year agriculture growth is less than 2 per cent. The question to debate is not what has gone wrong, what needs to be done to make our country free of poverty and hunger. Several challenges exist like land constraints, water shortage, climate change, low productivity. To handle this, we need to encourage the research and development in agriculture in a big way. If Israel can use their scanty rain filled desert land to produce oranges and tomatoes, why can't we do it? Yes, we can if we put our minds to it.
  - India ranks second highest worldwide in farm output
  - India is the largest producer of tea, mangoes, sugarcane, banana, turmeric, milk, coconut, pulses, ginger, cashew nuts, and black pepper.
  - India is the second highest producer of wheat, rice, sugar, vegetables, fruits and groundnut and cotton
  - India accounts for 10 per cent of the world's fruit production.

Chapter 3

Past and Present Status of Agriculture and Farmers

Status of Agriculture

In India agriculture is broadly classified in to five different periods before India's independence, which are as under.:

1. Early history (Before 15000 BCE)
2. Vedic period – Post Maha Janapadas period (1500 BCE – 200 CE)
3. Early Common Era – High Middle Ages (200–1200 CE)
4. Late Middle Ages – Early Modern Era (1200–1757 CE)
5. Colonial British Era (1757–1947 CE)

[Where: BCE - short for "Before the Common Era", "Before the Christian Era", or "Before the Current Era". CE - Common Era, Current Era (Christian Era is, however, also abbreviated AD, for Anno Domini]

1. Early History (Before 1500 BCE)

- ☆ *9000 BCE:* Wheat and barley were domesticated in the Indian subcontinent. Domestication of horse, sheep and goat soon followed. This period also saw the first domestication of the elephant.
- ☆ *8000-6000 BCE:* Barley and wheat cultivation, along with the domestication of cattle, primarily sheep and goat–was visible in Mehrgarh (Balochistan, now in Pakistan). Agro pastoralism in India included threshing, planting crops in rows–either of two or of six–and storing grain in granaries.

- ☆ *5000 BCE:* Agricultural communities became widespread in Kashmir.
- ☆ *5000-4000 BCE:* Cotton was cultivated. The Indus cotton industry was well developed and some methods used in cotton spinning and fabrication continued to be practiced till the modern Industrialization of India. A variety of tropical fruit such as mango and muskmelon are native to the Indian sub-continent. The Indians also domesticated hemp, which they used for a number of applications including making narcotics, fibre and oil. The farmers of the Indus Valley grew peas, sesame, and dates. Sugarcane was originally from tropical South Asia and Southeast Asia. Different species likely originated in different locations with *S. barberi* originating in India and *S. edule* and *S. officinarum* coming from New Guinea.
- ☆ *5440 BCE:* Wild Oryza rice appeared in the Belan and Ganges valley regions of northern India. Rice was cultivated in the Indus Valley civilization.
- ☆ *4500 BCE:* Irrigation was developed in the Indus Valley Civilization. The size and prosperity of the Indus civilization grew as a result of this innovation, which eventually led to more planned settlements making use of drainage.
- ☆ *3000 BCE:* Sophisticated irrigation and water storage systems were developed by the Indus Valley civilization, including artificial reservoirs at Girnar.
- ☆ *2600 BCE:* An early canal irrigation system from Circa.
- ☆ *2500 BCE:* Archeological evidence of an animal-drawn plough in the Indus Valley civilization
- ☆ *2000 BCE:* Agricultural activity included rice cultivation in the Kashmir and Harrappan regions.

2. Vedic Period-Post Maha Janapadas Period (1500 BCE–200 CE)

- ☆ Gupta (2004) finds it likely that summer monsoons may have been longer and may have contained moisture in excess than required for normal food production. One effect of this excessive moisture would have been to aid the winter monsoon rainfall required for winter crops.
- ☆ In India, both wheat and barley are held to be Rabi (winter) crops and–like other parts of the world–would have largely depended on winter monsoons before the irrigation became widespread. The growth of -the Kharif crops would have probably suffered as a result of excessive moisture.
- ☆ Jute was first cultivated in India, where it was used to make ropes and cordage.
- ☆ Some animals–thought by the Indians as being vital to their survival–came to be worshiped.

- Trees were also domesticated, worshiped, and venerated–Pipal and Banyan in particular.
- Others came to be known for their medicinal uses and found mention in the holistic medical system Ayurveda.
- *1000–500 BCE:* There are repeated references to iron. Cultivation of a wide range of cereals, vegetables and fruits is described. Meat and milk products were part of the diet; animal husbandry was important. The soil was ploughed several times. Seeds were broadcasted. Fallowing and a certain sequence of cropping were recommended. Cow dung provided the manure. Irrigation was practiced.
- *322–185 BCE:* The Mauryan Empire categorized soils and made meteorological observations for agricultural use. Other Mauryan facilitation included construction and maintenance of dams, and provision of horse-drawn chariots–quicker than traditional bullock carts.
- *300 BCE:* The Greek diplomat Megasthenes, in his book *Indika*– provides a secular eyewitness account of Indian agriculture.

3. Early Common Era–high Middle Ages (200–1200 CE)

- The Tamil people cultivated a wide range of crops such as rice, sugarcane, millets, black pepper, various grains, coconuts, beans, cotton, plantain, tamarind and sandalwood. Jackfruit, coconut, palm, areca and plantain trees were also known.
- Systematic ploughing, manuring, weeding, irrigation and crop protection was practiced for sustained agriculture. Water storage systems were designed during this period.
- Kallanai (1st-2nd century CE), a dam built on river Kaveri during this period, is considered the as one of the oldest water-regulation structures in the world still in use.
- Spice trade involving spices native to India–including cinnamon and black pepper–gained momentum as India starts shipping spices to the Mediterranean.
- Roman trade with India followed as detailed by the archaeological record and the *Periplus of the Erythraean Sea.*
- Chinese sericulture attracted Indian sailors during the early centuries of the Common Era.
- *320-550 CE:* Crystallized sugar was discovered by the time of the Guptas and the earliest reference of candied sugar come from India.
- *647 CE:* Chinese documents confirm at least two missions to India, initiated in, for obtaining technology for sugar-refining.
- *875-1279 CE:* Noboru Karashima's research of the agrarian society in South India during the Chola Empire reveals that during the Chola rule

land was transferred and collective holding of land by a group of people slowly gave way to individual plots of land, each with their own irrigation system.

- The growth of individual disposition of farming property may have led to a decrease in areas of dry cultivation.
- The Cholas also had bureaucrats which oversaw the distribution of water--particularly the distribution of water by tank-and-channel networks to the drier areas.

4. Late Middle Ages – Early Modern Era (1200–1757 CE)

- The construction of water works and aspects of water technology in India is described in Arabic and Persian works. The diffusion of Indian and Persian irrigation technologies gave rise to irrigation systems which bought about economic growth and growth of material culture.
- Agricultural 'zones' were broadly divided into those producing rice, wheat or millets.
- Rice production continued to dominate Gujarat and wheat dominated north and central India.
- The Encyclopædia Britannica details the many crops introduced to India during this period of extensive global discourse.
- *1556-1605 CE:* Land management was particularly strong during the regime of Akbar the Great under whom scholar-bureaucrat Todarmal formulated and implemented elaborated methods for agricultural management on a rational basis.
- Indian crops–such as cotton, sugar, and citric fruits–spread visibly throughout North Africa, Islamic Spain, and the Middle East.
- Though they may have been in cultivation prior to the solidification of Islam in India, their production was further improved as a result of this recent wave, which led to far-reaching economic outcomes for the regions involved.

5. Colonial British Era (1757–1947 CE)

- A number of irrigation canals are located on the Sutlej river.
- Few Indian commercial crops - such as Cotton, indigo, opium, and rice - made it to the global market under the British Raj in India.
- The second half of the 19th century saw some increase in land under cultivation and agricultural production expanded at an average rate of about 1 per cent per year by the later 19th century.
- Due to extensive irrigation by canal networks Punjab, Narmada valley, and Andhra Pradesh became centers of agrarian reforms.

- The British regime in India did supply the irrigation works but rarely on the scale required.
- Community effort and private investment soared as market for irrigation developed.
- Agricultural prices of some commodities rose to about three times between 1870-1920.
- A rich source of the state of Indian agriculture in the early British era is a report prepared by a British engineer, Thomas Barnard, and his Indian guide, Raja Chengalvaraya Mudaliar, around 1774. This report contains data of agricultural production in about 800 villages in the area around Chennai in the years 1762 to 1766. This report is available in Tamil in the form of palm leaf manuscripts at Thanjavur Tamil University, and in English in the Tamil Nadu State Archives.
- *1871:* Government of India created Department of Revenue, Agriculture and Commerce which formed as base for Initiation of Agriculture in India.
- *1880:* Famine Commission Report was submitted which was base for inception of Agricultural Department.
- *1881:* Separate Department of Agriculture at Centre for Famine relief operations.
- *1890:* Dr. J.A. Voelcker appointed as a consulting chemist from Royal Agricultural Society (England) - Laid foundation for agricultural research in India.
- *1892–1903* - Appointment of Imperial Agricultural Chemist, Imperial Mycologist and Imperial Entomologist – Base for Beginning of inducting the scientist in Agriculture.
- *1901-05:* To enhance agricultural education, Establishment of Agricultural Colleges at Pune, Kanpur, Sabour, Nagpur, Coimbatore and Lyallpur (Now in Pakistan).
- *1905:* Establishment of Imperial Agricultural Research Institute (IARI) at Pusa (Bihar).
- *1929:* Based on Royal Commission on Agriculture's recommendation (1928), Imperial Council of Agricultural Research (ICAR) was establishment to conduct comprehensive research.
- *1931-47:* Indian Lac Cess Committee, Indian Central Tobacco Committee, Indian Central Oilseeds Committee were formed to improve research in various crops.

Republic of India (1947 CE onwards)

- Special programs were undertaken to improve food and cash crops supply.
- The Grow More Food Campaign (1940s) and the Integrated Production Programme (1950s) focused on food and cash crops supply respectively.

- *1957:* All India Coordinated Maize Improvement Project was initiated (First coordinated project) to exploit maize research (Specifically heterosis).
- Five-year plans of India - oriented towards agricultural development - soon followed.
- *1963:* Introduction of semi dwarf wheat varieties from CIMMYT, Mexico Formed basis for green revolution.
- *1966:* Introduced semi-dwarf rice varieties TN1 and IR 8 from Taiwan and Philippines respectively is formed as base for green revolution.
  - Land reclamation, land development, mechanization, electrification, use of chemicals–fertilizers in particular, and development of agriculture oriented 'package approach' of taking a set of actions instead of promoting single aspect soon followed under government supervision.
  - The many 'production revolutions' initiated from 1960s onwards included Green Revolution in India, Yellow Revolution (oilseed: 1986-1990), Operation Flood (dairy: 1970-1996), and Blue Revolution (fishing: 1973- 2002) *etc.*
  - *1979:* National Agricultural Research Project (NARP) was launched to strengthen the research capabilities of SAUs.
  - Following the economic reforms of 1991, significant growth was registered in the agricultural sector, which was by now benefiting from the earlier reforms and the newer innovations of Agro-processing and Biotechnology.
  - *1998:* National Agricultural Technology Project (NATP) was initiated Strengthen the research on location specific problems Contract farming– which requires the farmers to produce crops for a company under contract–and high value agricultural product increased.
  - *2006:* National Agricultural Innovative Project (NAIP) was launched for End to end approach for solving problems.

Time Line of Development of Agriculture

1. 9500 BCE (Earliest evidence for domesticated wheat)
2. 9000 BCE (Early cultivation of plants and domestication of crops and animals)
3. 8000 BCE (Evidence for cattle herding)
4. 7000 BCE (Cultivation of barley; animals are domesticated)
5. 6500 BCE (Cattle domestication in Turkey)
6. 6000 BCE (Indus Valley grows from wheat to cotton and sugar)
7. 5500 BCE (Sumerians start organized agriculture)

8. 5400 BCE (Archaelogical proof for domestication of chicken)
9. 5400 BCE (Linearbandkeramik Culture in Europe)
10. 5000 BCE (Africa grows rice, sorghum)
11. 4000 BCE (Ploughs make an appearance in Mesopotamia)
12. 3000 BCE (Maize is domesticated in Americas)
13. 3000 BCE (Turmeric is harvested at Indus Valley).
14. 2737 BCE Tea is discovered
15. 2000 BCE 1st windmill in Babylon
16. 1000 BCE sugar processing in India
17. 500 BCE Row cultivation in China
18. Year 200 (Multi-tube seed drill invented in China)
19. Year 700 (Arab Agriculture Revolution)
20. Year 1000 (Coffee originates in Arabia)
21. Year 1492 (Columbian exchange changes agriculture)
22. Year 1599 1st Practical Greenhouse is created
23. Year 1700 (British Agricultural Revolution)
24. Year 1700 (Charles Townshend popularizes)
25. Year 1794 (Cotton gin is invented)
26. Year 1800 (Chemical fertilizer began to be used)
27. Year 1837 (John Deere invents steel plough)
28. Year 1860 (Hay cultivation changes)
29. Year 1866 (Gregor Mendel describes Mendelian inheritance)
30. Year 1879 (Milking machine replaces hand milking)
31. Year 1892 (First practical gasoline-powered tractor)
32. Year 1900 (Birth of industrial agriculture)
33. Year 1930 (First aerial photos for agriculture)
34. Year 1930 (First plant patent is given)
35. Year 1939 (DDT becomes a rage)
36. Year 1944 (Green Revolution begins in Mexico)
37. Year 1972 (Organic movement starts taking roots)
38. Year 1996 (Commercial cultivation of genetically modified plants)
39. Year 1998 National Agricultural Technology Project (NAPT) was initiated.
40. Year 2006 (National Agricultural Innovative Project (NAIP))
41. Year 2007 (National Food Security Mission)
42. Year 2009-10 (National Livelihood Mission)
43. Year 2014 (Pradhan Mantri Fasal Bima Yojana (PMFBY))

44. Year 2015 (Pradhan Mantri Gram Sinchai Yojana (PMGSY))
45. Year 2015 (Soil Health Card Scheme)
46. Year 2015 (Krishi Dak Service)
47. Year 2016 (DBT for Fertilizer Subsidy)
48. Year 2018 (Pradhan Mantri Kisan SAMPADA Yojana)

Status of Farmers in Society

Most of the Indian farmers are now leading a very miserable life. Industry has not thriven much in our country. Hence most of the people depend on agriculture. As a result there has been a great pressure on land. Each cultivator has got a very small amount of land. He gets only a small income from the produce of his land. So he cannot maintain his family with that income. For this reason he has to run into debt. Thus we see that the farmers as a class are debtors. We cannot describe what miserable life they lead. They live in wretched huts. They cannot always get two meals a day. They have not even ordinary clothes to wear. When they fall ill, they cannot meet the expenses of treatment and medicine. So, for want of food, drinking water, good dwelling houses they fall ill and die in large numbers for want of treatment and diet.

Measures to Improve Farmers Condition in India

- ☆ They are to be educated–The farmers are the backbone of the nation. So it is the first duty of the people and the Government of the country to improve their condition. If their condition is not improved, the whole nation will suffer. Generally farmers of our country are not educated. As they are poor, they cannot bear the educational expenses of their children. So they must be given free and compulsory primary education. Thus every cultivator must know how to read, write and do figure works. Then they must learn the principles of scientific agriculture.
- ☆ The old method of cultivation should be changed– Our farmers do not know how to cultivate the land scientifically. They follow the ancient method of tilling the land with a plough and a pair of oxen. But in Western countries, the farmers use motor tractors. Within a short time they till many acres of land and get more crops. But the Indian farmers cannot till so much land and do not get so much crops. Hence, they should be trained to use motor tractors to till the land. The Western farmers use improved manure to make the soil fertile. For this reason also they get more crops. Indian farmers generally use only cow-dung as manure. This also they do not use in large quantities. Hence the fertility of the soil is not increased and the land does not yield more crops.

- ☆ In Western countries the agricultural land is not divided into small plots. A farmer has got many acres of land in one big plot. It is economical to cultivate such a plot with tractors. In our country, the farmers have got small plots of land and the plots of a cultivator are situated at different places. Motor tractors cannot be used in such small plots. The cost of cultivation also is greater. Hence the farmers should join together and till their land with tractors. It is now done in Russia with great success. This is called collective farming. Like India, Russia also was mainly an agricultural country and the condition of the farmers was very miserable. But now scientific agriculture has been introduced there and the farmers are well off now and the country also has been prosperous.

India is mainly an agricultural country. Hence her prosperity depends largely on the improvement of agriculture. This can be done if the condition of the farmers can be improved. If agriculture is neglected, all classes of people will suffer. There had been, for some years, shortage of food crops in India. We had to buy food crops from other countries at a very high price. Hence, we all now feel the importance of growing more food. So for this reason at present there has been some improvement in this respect. More food crops are being produced now. As a result, our country is now self-sufficient in the matter of food crops. The farmers should get every facility to grow more food. If we can depend on our own food crops, we can save much money. This money may be spent in buying machinery for the development of industry. Considering all these, the Government has laid great importance on the problem of growing more food.

- ☆ They should learn thrift–Our farmers should learn to be thrifty. They spend more than what they earn. When they reap the harvest, they sell the crops and get money. At that time they spend money lavishly. They do not then think of the future. Within a short time they spend all their money. Then they are compelled to borrow money. In this way they run into debt.
- ☆ They should learn some craft–Indian farmers are not engaged for all the time in cultivating their land. Sometimes they have to remain idle. If they learn some handicrafts, such as spinning, weaving, they can earn some money and supplement their income from cultivation. They can thus be free from want.

Duty of the Government

- ☆ As the prosperity of the country depends largely on agriculture, our Government should try to improve the condition of the farmers.
- ☆ They should set up agricultural schools and model farms to educate the farmers free of cost. They should irrigate the land to make it fertile.

- ☆ They should lend money to the farmers at a nominal interest during the season of cultivation.
- ☆ They should distribute good seeds and manure to the farmers.
- ☆ The farmers may not be able to buy tractors.
- ☆ Our Government should lend tractors to them at a nominal charge.

Chapter 4

Astronomy: Prediction of Monsoon

Astronomy: An Overview

☆ Astronomy is the scientific study of celestial objects (such as stars, planets, comets, and galaxies) and phenomena that originate outside the Earth's atmosphere (such as the cosmic background radiation). It is concerned with the evolution, physics, chemistry, meteorology, and motion of celestial objects, as well as the formation and development of the universe. Astronomy is one of the oldest sciences.

General Signs that Bring Rain

- Soft, white, deep halo round the Moon or the Sun.
- Dark colored sky, dark as the crow's egg.
- Sky overcast with huge, bright, dense clouds.
- Needle-shaped clouds.
- Blood-red clouds.
- Rainbow in the morning or in the evening.
- Low, rumbling roar of thunder.
- Lightning.
- The appearance of the mock-sun; and
- Planets shine in full form and with soft light.

☆ Astronomers of early civilizations performed methodical observations of the night sky, and astronomical artifacts have been found from much earlier periods. However, the invention of the telescope was required before astronomy was able to develop into a modern science.

☆ Historically, astronomy has included disciplines as diverse as astrometry, celestial navigation, observational astronomy, the making of calendars, and even, at one time, astrology, but professional astronomy is nowadays often considered to be identical with astrophysics. Since the 20th century, the field of professional astronomy split into observational and theoretical branches.

- Observational astronomy is focused on acquiring and analyzing data, mainly using basic principles of physics.
- Theoretical astronomy is oriented towards the development of computer or analytical models to describe astronomical objects and phenomena.
- The two fields complement each other, with theoretical astronomy seeking to explain the observational results, and observations being used to confirm theoretical results.
- Amateur astronomers have contributed to many important astronomical discoveries, and astronomy is one of the few sciences where amateurs can still play an active role, especially in the discovery and observation of transient phenomena.

The Zodiac

- Zodiac is the division of the heavens into twelve astrology signs, each comprising exactly one-twelfth of the heavenly circle or 30° and totalling 360°. The Zodiac is a circle of space surrounding the Earth. It may be imagined as a belt in the heavens about 15 degrees wide in which the planets travel. It is the Sun's apparent path that is called ecliptic.
- The zodiacal circle is divided into twelve parts, each part containing thirty degrees of space called the signs of the Zodiac. In this circle the planets travel each in its own orbit, one outlaying beyond the other. The twelve signs of the Zodiac are Aries (Mesha), Taurus (Vrishabh), Gemini (Mithuna), Cancer (Kataka), Leo (Simha), Virgo (Kanya), Libra (Thula), Scorpio (Vrischika), Sagittarius (Dhanus), Capricorn (Makara), Aquarius (Kumbha) and Pisces (Meena).
- It is the twelve signs through which the planets travel or transit from west to east, going through one sign after another in their order from Aries to Pisces. Each sign possesses a specific influence. The planets also as they travel around the Zodiac exert an influence according to their separate nature and position in the Zodiac.
- According to modern Astrology there are twelve planets *viz*., Sun, Moon, Mars, Mercury, Jupiter, Venus, Saturn, Rahu, Kethu, Uranus, Neptune and Pluto, Hindu Astrology recognises only the first nine.
- **Quadruplicity:** Each sign belongs to one of four groups of signs based on their elemental tendencies to be either fiery, earthy, airy, or watery in temperament.

Elements	*Rasi/Zodiac/Signs*
Fire	Aries, Leo, Sgittarius
Earth	Taurus, Virgo, Capricorn
Air	Gemini, Libra, Aquarius
Water	Cancer, Scorpio, Pisces

Note: Capricorn is half-watery and half-earthy.

Seasons and Equinoxes

- Sun, the latter's rays fall equally only on two days in a year *i.e.* the day and nights are equal on two days a year when the Sun enters the Equator. These two days are March 21st and September 21st. One is called autumnal equinox and the other is called vernal equinox.
- These points are also known as equinoctical points or Vishu bindus. There are basically six seasons. Each of the above six seasons has been divided into two parts.

Seasons	*Parts*	*Period*
Vansantha Ritu	Madhu	March 21st to April 21st
	Madhava	April 21st to May 21st
Greehma Ritu	Sukra	May 21st to June 21st
	Suchi	June 21st to July 21st
Varsha Ritu	Nabhas	July 21st to August 21st
	Nabhasya	August 21st to September 21st
Shard Ritu	Lsa	September 21st to October 21st
	Urija	October 21st to November 21st
Hemantha Ritu	Sahas	November 21st to December 21st
	Sahasya	December 21st to January 21st
Sisira Ritu	Tapas	January 21st to February 21st
	Tapasya	February 21st to March 21st

- The Moon which is a natural satellite of the Earth moves around the Earth once in about 28 days, *i.e.* it takes about 28 days to come to the some star, after going around the earth. This is called the Sidereal Movement of the Moon.
- There is another way of recognizing the movement of the Moon around the Earth and that is with respect to the Sun. On Amavasya (New Moon) Day Poornima (Full Moon) Day, the Sun, the Moon, the Earth are on the same line longitudinally. From one Amavasya (New Moon) to another Amavasya (New Moon) it takes about 30 days.

- Each of these divisions (of 30) is called a Thithi (phase). This division of months is called Luni-solar Months. This is recognized in the Panchanga (Caiendar) as Prathma (1st day), Dwithiya (2nd day) *etc.* Sukla prathama (1st day of Bright half) or Krishna Prathama *etc.* (1st day of Dark half) depending on the bright or dark fortnight.

Sukala Paksham/Krishna Paksham

Sl.No.	*Sukala Paksham Thithi*	*Sl.No.*	*Krishna Paksham Thithi*
1	Prathama	1	Prathama
2	Dwithiya	2	Dwithiya
3	Thrityiya	3	Thrityiya
4	Chaturthi	4	Chaturthi
5	Panchami	5	Panchami
6	Sashti	6	Sashti
7	Sapthami	7	Sapthami
8	Ashtami	8	Ashtami
9	Navami	9	Navami
10	Dasami	10	Dasami
11	Ekaadasi	11	Ekaadasi
12	Dwaadasi	12	Dwaadasi
13	Thryodasi	13	Thryodasi
14	Chaturidasi	14	Chaturidasi
15	Poornima or Full Moon	15	Amavasya or new moon

- **Planetary movement:** There are five planets moving around the sun. They are Mercury, Venus, Mars, Jupiter and Saturn. By the time mercury, goes round the sun, approximately 88 days. Would have elapsed (4 rounds a year); by the time venus moves round the sun, it will be 224 days; for mars it is 686 days, for Jupiter 4332 days; and for Saturn 10,759 days.
- **Ancient systems of time:** Our ancient calculated time is an follows : one day = 60 Naligais 1 Naligai =60 vikalas. Therefore a day = 60 x 60 x 60 x vikalas = 2,16,000. The Rig veda contains 4,32,000 units of sounds therefore 1 vikala = 2 units of sound.

The ancient/indigenous methods of weather forecast may be broadly classified into two categories:

(i) Observational Methods

- Atmospheric changes
- Bio-indicators
- Chemical changes

- ☆ Physical changes
- ☆ Cloud forms and other sky features

(ii) Theoretical Methods or Astrological Factors or Planetary Factors

- ☆ Computation of planetary positions and conjunctions of plants and stars
- ☆ Study of solar ingress and particulars dates of months
- ☆ Study of Nakshatra Chakras
- ☆ Study of Nadi Chakras
- ☆ Dashatapa Siddhana

Prediction of Monsoon Rains

- ☆ **Parashara's technique** of *'rain forecast'* is based on the positions of the Sun and the Moon.

Sign of the Sun	*Sign of Moon*	*Predicted Annual Rainfall*
Cancer	Gemini, Aries, Taurus, or Pisces	100 adhakas
Leo or Sagittarius	Gemini, Aries	50 adhakas
Virgo or Leo	Gemini, Aries, Taurus or Pisces	80 adhakas
Cancer, Aquarius, Scorpio, or Libra	Gemini Aries, Taurus or Pisces	96 adhakas

Parasara's Rainfall Prediction

- ☆ Every year has (a particular planet as) a ruler, (another planet as) a minister, a particular cloud, and (depending on that) an amount of rainfall which one has to study to acquire the knowledge of rains. The method of finding out the ruler (planet) of the year: Multiply the number denoting the Saka year by three. Add two; divide the result by the number of sages (*i.e.*, seven). The remainder is the number indicating the ruling planet of that Saka year. The planet, which is fifth form the ruler planet, indicates the minister planet of that year. The minister plant of the year is Venus as it is the fifth from the Sun.

  1920 x 3 = 5760

  5760 + 2 = 5762

  5762/7 = 823+ remainder 1

- ☆ The Sun as the ruler of the year indicates average rainfall, the Moon heavy rains, Mars scanty rains, and Mercury goods rains. When Jupiter happens to be the king of the year the rainfall is satisfactory, Venus indicates excellent rainfall while Saturn as a king leaves the earth dry and dusty. Diseases of the eye, threat of fever, and all sorts of other calamities, scanty rainfall and continuous blowing of winds are the characteristics of a year ruled by the Sun. The year in which the Moon is the ruler is sure to enrich the earth with good harvest and bestow health on mankind. In the year ruled by Mars, damage is caused to the crops and diseases spread among people. The earth becomes benefit of crops. When Mercury happens to be the ruler, earth is free of diseases. Transportation is easy and there is plenty of harvest. The earth is blessed with all the varieties of crops. If Jupiter rules the year, Dharma prevails on earth, people enjoy peace of mind, There is good rainfall The whole earth enjoys prosperity. Venus the preceptor of demons, as a ruler of the year causes the kings to prosper without fail. Prosperity and plenty result. The earth is blessed with a variety of food grains. The year in which Saturn rules war, stormy rains and outburst of diseases are sure to occur. Rains rare scanty and winds are continuous.

Table 4.1: Annual Rainfall and Crop Yields Depending on the Ruling Planets of the Year

Name of the Ruling Plant of the Year	*Estimated Rainfall for the Year*	*Crop Yield during the Year*
Sun	Average or scanty	Poor crop yield
Moon	Heavy	Good harvest
Mars	Scanty	Damage to crops
Mercury	Good	Plenty of harvest
Jupiter	Satisfactory	Good harvest[1]
Venus	Excellent	Variety of food grains
Saturn *	Scanty	Poor yield

*Saturn: The earth is dry and dusty, continuous winds occur during this period.

A model for forecasting seasonal rainfall recorded in Brhat Samhita

- Varahamihira (600 AD) evolved or adapted a technique based on science. This techniques lays down that after the occurrence of the full-moon day of the month of Jyestha (approximately coinciding with June of Gregorian calendar) the asterism or lunar mansion or naksatra of the day on which the first rainfall of that yeara rainy season is received should be noted. This asterism provided the basic for the forecast of seasonal rains. The predicted amount of the season's total rainfall for each nakshatra or lunar mansion if it happens to be the nakshatra on the first rainfall of the season is listed (Table 4.2). The first rainfall of the season that occurred after the full-moon day of the month of Jyestha (approimately June) is taken into account for forecasting the seasonal rainfall, but the amount of rainfall recorded on that day has hot been indicated. Modern meteorology defines a rainy day as a day on which a rainfall of 2.5 mm or more has been recorded.

Table 4.2: Varahamihira's Technique for Forecasting Seasonal Rains

Lunar Mansion	*Zodiac Sign*		*Predicted Total Seasonal Rainfall*	
	Sanskrit	*English*	*In Ancient Units (Drones)*	*In Modern Units (cm)*
Hasta	*Kanya*	Virgo	16	102.4
Purvashadha	*Dhanu*	Sagittarius	16	102.4
Mrigashirsha	*Vrushabha*	Taurus	16	102.4
Chitra	*Kanya*	Virgo	16	102.4
Revati	*Meena*	Pisces	16	102.4
Dhantishtha	*Makara*	Capricorn	16	102.4
Shatabhisha	*Kumbha*	Aquarius	4	25.6
Jyeshtha	*Vrushchika*	Scorpio	4	25.6
Swati	*Tula*	Libra	4	25.6
Kritika	*Vrushabha*	Taurus	10	64.0
Shravana	*Makara*	Capricorn	14	89.6
Magha	*Simla*	Leo	14	89.6
Anuradha	*Vrushchika*	Scorpio	14	89.6
Bharani	*Mesha*	Aries	14	89.6
Mula	*Dhanu*	Sagittarius	14	89.6
Purvaphalguni	*Simla*	Leo	25	160.0
Punarvasa	*Mithun*	Gemini	20	128.0
Vishakha	*Vrushchika*	Scorpio	20	128.0
Uttarashadha	*Makara*	Capricorn	20	128.0
Aaslesha	*Karka*	Cancer	13	83.2
Uttarabhadrapada	*Meena*	Pisces	25	160.0
Uttaraphalguni	*Kanya*	Virgo	25	160.0
Rohini	*Vrushabha*	Taurus	25	160.0
Purvabhadrapada	*Kumbha*	Aquarius	15	96.0
Pushya	*karka*	Cancer	15	96.0
Ashwini	*Mesha*	Aries	12	76.8
Aradra	*Mithun*	Gemini	18	115.2

On the day of the first rainfall of the season I drona=6.4cm.

The Method of Ascertaining the Type of Cloud of the Year

☆ Add the types of fire (which is three) to the number denoting the Saka year. Divide the sum by the number of vedas (which is four). The remainder of the division indicates the type of cloud, *viz.*, Aavarta, *etc.*, according to their order. Let the Saka year is 1920. The 1920+3 = 1923; 1923/4 = 480+ remainder 3. Hence the type of cloud is the one listed at number 3.

Pushkara cloud is stated at number 3 in the order. Therefore the cloud of the Saka year 1920 is Pushkara. Aavarta, Samvarta, Pushkara, and Drona are the four types of clouds, Aavarta being the first in order.

The Method of Determining the Amount of Rainfall

- Experts have fixed *adhaka* (jalaadhaka) as the measure of water which is the quantity of water contained in an expanse of a hundred *yojanas* and the depth of thirty *yojanas*. When the Sun enters the sign of cancer while the moon is in Pisces, Aries, Taurus or Gemini, the rainfall is a hundred adhakas.
- If the sun passes through Leo and Sagittarius it is half of that (*i.e.*, fifty adhakas). when the sun is in the Virgo or Leo rainfall is stated to be eighty adhakas. When the Sun is in Cancer, Libra, Aquarius, the rainfall is said to be ninety-six adhakas. Farming should be planned after studying the quantity of rainwater.

Sudden Rainfall

- If an expert on predictions of rainfall is approached with query regarding rains while he is taking a dip in water or has water in his hand or is the variety of water, sudden rains can be predicted:
  - Ants emerging (from the ant hill) carrying their eggs and a sudden croaking of frogs are also indications of sudden rains.
  - Cats, mongooses, snakes, other creatures which live in holes as well as grasshoppers moving around freely as in a state intoxication are also sure signs of sudden rains.
  - Children playing on the road and building bridges of mud, and peacocks dancing also indicates sudden rains without fail.
  - People suffering from injury or vatadosha (human disorder similar to wind humor) complaining of body pain and snakes climbing on the treetops also bespeak of sudden rains.
  - Water birds drying their wings in the hot sun and crickets chirping in the sky also signify sudden rains.

Indications of Famine

- Mar's transit through Dhruva (Uttaraphalguni, Uttarashadha and Uttarabhadrapada nakshadars), Vaishanava (Shravana), Hasta, Mula, Shakra (the master of Jyeshtha), Krittika, and Magha indicates famine. The sun situated behind Mars evaporates even the ocean while in an opposite situation, he drenches mountains too.
- Obstruction to rain as soon as Venus reaches the middle of its path through Chitra. Mars passing through Leo turns the earth into a fireplace, and accompanied by the Sun can evaporate even the ocean.

Kautilya's Arthashastra (400 BC)

- ✰ He describes the technique for measuring rainfall for a location. A circular vessel with a diameter equal to the length of human arm (which is equal to the distance measured by the width of twenty fingers of a human hand) and a depth equal to the distance measured by the width of eight fingers (in modern unit, the diameter and the depth would approximately 38 cm and 13 cm respectively) was used to collect the rainwater. When this vessel was filled with rainwater collected open space, rainfall was measured to be 50 palas or one adhaka or ¼ drona.
- ✰ An adhaka of rainfall is equal to 1.6 cm rainfall in modern units of measurement and a drona is 6.4 cm.

Planets and its Role on Occurrence of Rain, Flood, Drought and Famine

- ✰ When sun is between Venus and Mercury there is a break in monsoon in the sense that for some days there is dry spell.
- ✰ Sun being behind Mars in the rainy season, there will be poor rain or rain is delayed or will create dry spells. When the Sun was overtaking Mars, there will be heavy downpour of rains, causing flood in rivers.
- ✰ Rain will not be timely when all quadrants being occupied by malefics.
- ✰ Mars, affected by other malefics, will create dry spells till August.

Some Animal Behaviour and Forecasts

- ❖ In the rainy season when the sky is cloudy try to take your pet dog outdoor. If the dog shows a disinclination, it is a sign of coming rain.
- ❖ See if any spider has started weaving its web outdoors. It indicates the departure of the monsoon.
- ❖ Those who are lucky to have some frogs alive and croaking can get the indication from their croaking.
- ❖ The exultant cry of the peacock is an indication of cloud formation.
- ❖ Early flowing of the gulmohur and amaltas was an indication of a good monsoon.

- ✰ If Jupiter and Mars are within 30 degrees (thirty degrees) of each other it prevents rains.
- ✰ If the Moon is in the 7th from Venus and within view of benefic planets, or be in the 5th, 7th or 9th house from Saturn there will be immediate rain.
- ✰ Clouds become scattered and rainfall disturbed, when the sun, Mars and Venus transit the same sign. If Jupiter joins the above combinations, clouds will deliver rains in plenty.
- ✰ When Jupiter retrogrades in Rohini, the year will have less rainfall.
- ✰ Heavy rain results when Jupiter is in Pisces while Venus is in Cancer.

- Droughts are noticed when Saturn is unaspected in Aries, Leo or Sagittarius.
- When Mars and Saturn are in conjunction, rainfall will be very low.
- When Venus is in constellations of Swathi, Vishakha and Anusha, unprecececedent rainfall results in heavy floods.
- Famine will break out for want of rains when Venus is in one of constellations from Jyestha to Sravana.
- There will be drought condition when Venus sets in or retrogrades in makha or Uttarashadha.

Rain Forecasting in Indian Panchangs

- An almanac is a book or table containing a calendar of the days, weeks and months of the year, a register of ecclesiastical festivals ad saint's days and a record of various astronomical phenomena, often with weather prognostications and seasonal suggestions for the countrymen.
- In India, the classical Hindu almanac is known as 'Panchang'. For astrologers, it is one of' basic books for making astrological calculations, casting horoscopes, and for making predictions. For farmers, it is an astrological guide to start any farming activity. Hence, it is a fundamental book which is referred to by a large section of the people in this country various purposes.
- The word 'Panchang' has its roots in two Sanskrit words, *viz.*, 'panch' and 'ang', which mean 'five' and 'bodypart/limb' respectively. These parts are:
  - Tithi or lunar day-there are a total of thirty tithis in a lunar month, fifteen in each fortnight;
  - Vara or week day-' there are seven varas, namely, Ravivara (Sunday), Somavara (Monday), Mangalavara (Tuesday), Budhavara (Wednesday), Guruvara (Thursday), Shukravara (Friday), and Shanivara (Saturday);
  - Nakshatra or asterism or constellation – there are a total of twenty seven nakshtras named according to the yogataras or identifying stars of each of the twenty seven equal parts of the ecliptic or solar path;
  - Yoga or time during which the joint motion of the Sun and the Moon covers the space of a nakshatra, and
  - Karana or half of a lunar day or half-tithi.

Making Krishi-Panchang

- A *Krishi-Panchang* may be defined as "basic astro-agricultural guide book/calendar (that needs to be) published annually, giving calendrical information on various aspects of agricultural and allied activities, basically suggesting region-wise, season-wise, and crop-wise crop strategy based

on astro-meteorological predictions, giving auspicious/inauspicious time for undertaking/avoiding various farm-related operations, along with a list for performing religious rites, festivals, observing fasts, and some non-astrological agricultural guidance, primarily useful for the farming communities and persons having interest in agricultural development".

Content and Coverage Proposed

- ☆ The *Krishi-Panchang* should be basically different from the present-day *panchangs* in its content and coverage, method and approach of writing, composition of editorial boards, publication, and circulation. The *Krishi-Panchang*, being meant for meeting agricultural purposes, majority of its contents should relate to agricultural information. In addition to this, basic information such as annual date calendar, list of holidays, auspicious days/moments of the coming year should be given for the benefit of farming communities.
- ☆ The contents of the proposed *Krishi-Panchang* can broadly be categorized in two major groups as follows:

1. *Information which changes every year*

i. Annual date and Holiday calendar
ii. Month-wise daily guide for the whole year
iii. "*Rashiphal*", *i.e.*, month-wise forecasting of persons having different zodiac sings
iv. Daily/monthly/annual weather forecasting for the particular year
v. Crop prospects of that year based on planetary positions
vi. Season-wise crop strategy based on anticipated weather

2. *Information which remains same irrespective of any particular year*

i. Theories relating to agricultural and meteorological forecasting
ii. Auspicious moments for agricultural and allied activities
iii. Some general agricultural guidance

Modern Techniques of Weather Forecast

Persistence

- ☆ The simplest method of forecasting the weather, persistence, relies upon today's conditions to forecast the conditions tomorrow. This can be a valid way of forecasting the weather when it is in a steady state, such as during the summer season in the tropics.
- ☆ This method of forecasting strongly depends upon the presence of a stagnant weather pattern. Therefore, when in a fluctuating weather

pattern, this method of forecasting becomes inaccurate. It can be useful in both short range forecasts and long range forecasts.

Use of a Barometer

- Measurements of barometric pressure and the pressure tendency (the change of pressure over time) have been used in forecasting since the late 19th century. The larger the change in pressure, especially if more than 3.5 hPa (2.6 mmHg), the larger the change in weather can be expected.
- If the pressure drop is rapid, a low pressure system is approaching, and there is a greater chance of rain. Rapid pressure rises are associated with improving weather conditions, such as clearing skies.

Looking at the Sky

- Marestail shows moisture at high altitude, signalling the later arrival of wet weather. Along with pressure tendency, the condition of the sky is one of the more important parameters used to forecast weather in mountainous areas. Thickening of cloud cover or the invasion of a higher cloud deck is indicative of rain in the near future. High thin cirrostratus clouds can create halos around the sun or moon, which indicates an approach of a warm front and its associated rain.
- Morning fog portends fair conditions, as rainy conditions are preceded by wind or clouds that prevent fog formation. The approach of a line of thunderstorms could indicate the approach of a cold front. Cloud-free skies are indicative of fair weather for the near future. A bar can indicate a coming tropical cyclone. The use of sky cover in weather prediction has led to various weather lore over the centuries.

Nowcasting

- The forecasting of the weather within the next six hours is often referred to as nowcasting. In this time range it is possible to forecast smaller features such as individual showers and thunderstorms with reasonable accuracy, as well as other features too small to be resolved by a computer model.
- A human given the latest radar, satellite and observational data will be able to make a better analysis of the small scale features present and so will be able to make a more accurate forecast for the following few hours. However, there are now expert systems using those data and mesoscale numerical model to make better extrapolation, including evolution of those features in time.

Use of Forecast Models

- An example of 500 mbar geopotential height prediction from a numerical weather prediction model. In the past, the human forecaster was

responsible for generating the entire weather forecast based upon available observations.

- Today, human input is generally confined to choosing a model based on various parameters, such as model biases and performance. Using a consensus of forecast models, as well as ensemble members of the various models, can help reduce forecast error. However, regardless how small the average error becomes with any individual system, large errors within any particular piece of guidance are still possible on any given model run.
- Humans are required to interpret the model data into weather forecasts that are understandable to the end user. Humans can use knowledge of local effects that may be too small in size to be resolved by the model to add information to the forecast. While increasing accuracy of forecast models implies that humans may no longer be needed in the forecast process at some point in the future, there is currently still a need for human intervention.

Analog Technique

- The analog technique is a complex way of making a forecast, requiring the forecaster to remember a previous weather event that is expected to be mimicked by an upcoming event. What makes it a difficult technique to use is that there is rarely a perfect analog for an event in the future. Some call this type of forecasting pattern recognition. It remains a useful method of observing rainfall over data voids such as oceans, as well as the forecasting of precipitation amounts and distribution in the future.
- A similar technique is used in medium range forecasting, which is known as teleconnections, when systems in other locations are used to help pin down the location of another system within the surrounding regime. An example of teleconnections are by using *El Niño*-Southern Oscillation (ENSO) related phenomena.

Communicating Forecasts to the Public

- Most end users of forecasts are members of the general public. Thunderstorms can create strong winds and dangerous lightning strikes that can lead to deaths, power outages, and widespread hail damage. Heavy snow or rain can bring transportation and commerce to a stand-still, as well as cause flooding in low-lying areas. Excessive heat or cold waves can sicken or kill those with inadequate utilities, and droughts can impact water usage and destroy vegetation.
- Several countries employ government agencies to provide forecasts and watches/warnings/advisories to the public in order to protect life and property and maintain commercial interests. Knowledge of what the end user needs from a weather forecast must be taken into account to present the information in a useful and understandable way.

- ☆ Examples include the National Oceanic and Atmospheric Administration's National Weather Service (NWS) and Environment Canada's Meteorological Service (MSC). Traditionally, newspaper, television, and radio have been the primary outlets for presenting weather forecast information to the public. In addition, some cities had weather beacons. Increasingly, the internet is being used due to the vast amount of specific information that can be found. In all cases, these outlets update their forecasts on a regular basis.

Low Temperature Forecast

- ☆ The low temperature forecast for the current day is calculated using the lowest temperature found between 7 pm that evening through 7 am the following morning. So, in short, today's forecasted low is most likely tomorrow's low temperature.

Agriculture

- ☆ Farmers rely on weather forecasts to decide what work to do on any particular day. For example, drying hay is only feasible in dry weather. Prolonged periods of dryness can ruin cotton, wheat, and corn crops. While corn crops can be ruined by drought, their dried remains can be used as a cattle feed substitute in the form of silage. Frosts and freezes play havoc with crops both during the spring and fall.
- ☆ For example, peach trees in full bloom can have their potential peach crop decimated by a spring freeze. Orange groves can suffer significant damage during frosts and freezes, regardless of their timing.

Forestry

- ☆ Weather forecasting of wind, precipitations and humidity is essential for preventing and controlling wildfires. Different indices, like the Forest fire weather index and the Haines Index, have been developed to predict the areas more at risk to experience fire from natural or human causes.
- ☆ Conditions for the development of harmful insects can be predicted by forecasting the evolution of weather, too.

Chapter 5

Ancient Soil Classification and Soil Fertility

Introduction

- ☆ Soils exist in a very dynamic state purely in open system, wherein exact taxonomic grouping is not very simple, if not impossible, though efforts are being made for unified soil classification as well as universal soil classification. Not only the established soil forming factors are involved, but so many other factors do cause changes in soil unit (*pedon*). If soil erosion can cause degradation, management in some cases are improving soil's overall features.
- ☆ India is a country of distinctly varied climates, geology, relief and vegetation and thus possessing the distinct varieties of soil groups logically different from one another. Criteria used to classify the Indian soils are primarily based on the geology, relief, morphological features, chemical composition, physical structure and fertility.

Ancient Times and Early History

- ☆ A true soil science was not formed during ancient times and early history. Human Knowledge of soil was based on observation not on experimentation and testing of theories. Despites this much was learned about soils. In places all over the world various people discovered irrigation, basic erosion controls such a terracing the value of the plow in preparing the soil for planting, and recognized that some soils were more productive than others.

- ☆ In some places humans had even discovered how to create artificial soil or to improve soils not well suited to crops. Therefore, a solid knowledge base of soils as an agricultural medium had already been formed.

Soil Science in the Middle Ages (5th to 14th centuries AD)

- ☆ In western societies the Middle ages represented a period of repression for science and a temporary loss of soils knowledge held by groups such as the Greeks and Romans. This loss was due in large part to the dominance of religion in western life. However, soils knowledge still existed in this period and in some parts of the world was expanded.

Byzantium and Europe

- ☆ Following the sacking of Rome in 410 AD, Roman culture shifted to Byzantium in modem Turkey. Many Roman manuscripts, including agricultural manuscripts. Were moved to Byzantium. Therefore, many of the scientific ideas developed by the Roman Empire were preserved and advanced over the next 1000 year's by the Byzantines soils knowledge. It includes works by Roman soils specialists, but many new Byzantine authors are also represented. Important information included a description of the soils of the byzantine Empire, discussions of which crops were most appropriate for different soils, and ways to evaluate the quality of soils.
- ☆ Agriculture declined in Europe following the fall of the Roman Empire, a decline that included both the area of ;and under cultivation and the yields. Obtained during crop growth. Brief periods of renewed interest occurred in the 8th and 9th century. Draining of marshland, fertilization of the soil all helped to increase agricultural yields considerably manure was in fact a highly valued commodity due to its fertilizing abilities. Cultivated land on steep slopes were returned to forest as early as the 10th century AD in an effort to reduce soil erosion.

Arabia and the Middle East

- ☆ During the Middle Ages Islamic- based societies which had formed and spread form the Arabian peninsula, were among the world's leaders in science, math, and technology. This included the agricultural sciences. Earlier works form civilizations such as the Greeks. Romans, Chinese, and Indians were known to Muslim scientists, who studied, combined and built upon these earlier works.
- ☆ In agriculture, one hallmark of middle Ages Muslim government was the development and support of extensive networks of irrigation canals. Advanced Muslim mathematics contribute greatly to the engineering of these irrigation systems.

☆ Muslim agronomists were also adept at identifying soils suitable to the crops being grown. Libraries in major Muslim cities typically contained numerous agricultural calendar in the 10th century that listed, among other items, monthly tasks related to the preparation of soils for agriculture. Soil fertility was maintained through the use fertilizers such as manure and it was recognized that different crops had different fertility requirements.

Southeast Asia

☆ Southeast Asia was a region in which soils knowledge expanded during the Middle Ages. Chinese governments documents form this period separated soils into 12 categories based on the crops they were most suited to and regulated the time of working the fields. In each community and individual was responsible to looks after fertilizers and soil fertility. Terracing of farm fields in china for erosion control can be documented as early as the 7th century AD. The decrees of Chinese emperors showed a strong appreciation for soils. Emperor Ming ordered that all lands be divided based on their location and soils in 1387, and land surveys, including soils information, were made for large portions of the country.

☆ Japanese agriculture was influenced by the chines until the 9th century AD, after which the Japanese halted immigration and moved away from chinses influence. The lack of good land led the Japanese to Place a high value on soil. Many Forms of maintaining soil fertility, including manure green manure the growth of legumes, and crop rotations were used. Terraces were used on steep slopes. Artificial soils were created and land surveys were common. In India irrigation fertilization through manure application and fallow periods to restore fertility widely used by the 14th century ADand land surveys including soils were made for many parts of the country.

Ancient Soil Classification

☆ A number of literary works of the ancient period have mentioned. In ancient times geographical distribution by Surapala was jangala (arid), anupa (marshy) and samanya (ordinary). It is further divided by colour into black, white, pale, dark, red and yellow by taste into sweet, sour, salty, pungent, bitter and astringent. Samanya land was suitable for all kinds of trees.

☆ Rig-veda identified productive and non-productive soils. There were 12 types of land depending upon the fertility of soil irrigation and physical properties. The 12 types of soil are given below:

1. Urvara (fertile)
2. Ushara (barren)

3. Maru (desert)
4. Aprahata (fallow)
5. Shadvala (grassy)
6. Pankikala (muddy)
7. Jalaprayah (water)
8. Kachchaha (land contiguous to water)
9. Sharkara (full of pebbles)
10. Sharkaravari (sandy)
11. Nadimatruka (land water from river)
12. Devamatruka (rainfed)

Classification Based on Crops Suitability

According to crop suitability soils are countified as under:

1. Vrdiheyam (rice (rainfed)/corn)
2. Shaleyam (kamala (wet) rice)
3. Tilyam (sesamum)
4. Mashyam (blackgram)
5. Maudginam (mung bean)

Use of Manure in Ancient Times

- ✰ The ***Brhatsamhita*** prescribes that seeds which have been properly treated are to be sown with the addition of pork or venison into the soil and sprinkled daily with water mixed with milk (***ksira***). It says further: **'To promote inflorescence and fructification**, a mixture of one ***adhaka*** (64 palas) of sesame, 2 *adhaka* of excreta of goats or sheep, one ***prastha*** (16 palas) of barley powder, one ***tola*** of beef thrown into one ***drona*** (256 palas) of water and standing over seven nights should be poured round the roots of the plant.'
- ✰ In order to ensure sprouting and growth of luxurious stem and foliage, according to the ***Brhatsamhita*** the seed should be soaked in an infusion made of paddy powder, urad (masa), sesame and barley which are mixed with decomposing flesh and the whole mass steamed with the addition of **turmeric** (*haridra*).
- ✰ For the growth of ***kapittha*** (Feronia elephantum), the seeds should be soaked for a short time in a decoction of ***asphota*** (Jasmine), ***amalaki*** (*Phyllanthus embellicus*), ***dhava*** (Gris/eatomentosa), ***vasaka*** (*Justica adhatoda*), **vetula** (*Calamus rotung*), **suryavalli** (*Gynandropsis pentaphyta*), ***syama*** (*Echites fructescens*) and ***atimuktaka***(*Aganosma caryophyllata*) **boiled in milk**. The soaked seeds must be dried in the sun and process is to be repeated for a month. A circular hole is to be

made in the ground (1 cubit in diameter and 2 cubits in depth), and the milky decoction poured into it. When it dries up, it is burnt and pasted over with ashes mixed with ghee and honey. Three inches of soil should now be thrown into them along with the powder of bean, sesame and barley, and then again three inches of soil. Finally, washings of fish are to be sprinkled and the mud beaten to a thick consistency. Now the treated seeds should be placed in the hole.

- ☆ According to the ***Agnipurana*** tree becomes laden with flowers and fruits by manuring the soil with powdered barley, sesame and the offal matter of a goat mixed together and soaked in washings of beef for seven consecutive nights. A good growth of these is secured by sprinkling the washings of fish on them.
- ☆ The ***krsi Parasara*** says: 'In the month of Magha a dung heap is to be raised with the help of a spade. When it becomes dried in the sun, smaller balls are made out of it. In the month of *Phalguna*, the dried balls are placed into holes dug for the purpose in the field, and at the time of sowing they are scattered into the filed. The practice of not disturbing the dung heap for a month means the minimization of the loss of nitrogen, the chief fertilizing element, that of drying into balls results in the reduction of active ammonia which may be injurious to the plants, while that of placing the dung balls into pits increases the humus which undoubtedly contributes to the fertility of the soil.
- ☆ The use of flesh of animals, fish-washings, vegetable products and the farm-yard manure consisting of the excreta of various animals, mixed with litter which would absorb the animal urine, indicates that the farmers of the ancient period were well aware of their fertilizing property and also the physical effects of the manure upon the texture and water-holding power of the soil, although they did not certainly possess the precise chemical knowledge of these fertilizers. It is now known that the farm-yard manure contains all the essential plant nutrients, *viz.* nitrogen, phosphoric acid and potash.
- ☆ In the next part of this series we will understand in details about accessories, irrigation, crops, and treatment of plant diseases *etc.* practises of agriculture followed in classical age.

Maintenance of Soil Fertility

Soil Fertility

"The capability of the soil to provide all the essential plant nutrients in available form is called as soil fertility".

"The inherent capacity of the soil to supply nutrients to plants in adequate amount and in suitable proportions is called as soil fertility".

Types of Soil Fertility

1) Inherent or Natural Fertility

- ☆ The soil, as a nature contains some nutrients, which is known as inherent fertility.
- ☆ Among plant nutrients nitrogen, phosphorus and potassium is essential for the normal growth and yield of crop. The inherent fertility has a limiting factor from which the fertility is not decreased.

2) Acquired Fertility

- ☆ The fertility developed by application of manures and fertilizers, tillage, irrigation, *etc.*, is known as acquired fertility.
- ☆ The acquired fertility has also a limiting factor. It is found by experiment that the yield does not increase remarkably by application of additional quantity of fertilizers.

Factors Effecting Soil Fertility

The factors that are affecting soil fertility may be of two types:

(i) Natural Factors

The *natural factors* are those which influence the soil formation and the artificial factors are related to the proper use of land.

(ii) Artificial Factors

The factors effecting the fertility of soil are parent material, climate and vegetation, topography, inherent capacity of soil to supply nutrient, physical condition of soil, soil age, micro-organisms, availability of plant nutrients, soil composition, organic matter, soil erosion, cropping system and favourable environment for root growth.

Measures for Maintenance of Soil Fertility

Maintenance of soil fertility is a great problem of our farmers. Cultivation of particular crop year after year in the same field decreases the soil fertility. To increase the soil fertility, it is necessary to check the loss of nutrient and to increase the nutrient content of soil.

The following things must be properly followed for increasing the fertility of soil.

1. Proper use of land,
2. Good tillage,
3. Crop rotation,
4. Control of weeds,
5. Maintenance of optimum moisture in the soil,

6. Control of soil erosion,
7. Cultivation of green manure crops,
8. Application of manures,
9. Cultivation of cover crops,
10. Removal of excess water (drainage),
11. Application of fertilizers,
12. Maintenance of proper soil reaction.

Maintenance of Soil Productivity

Several traditional methods of maintenance of soil productivity are described as under:

- ✰ Traditional soil management practices are the product of centuries of accumulated knowledge, experience and wisdom refined and perpetuated over generations. These practices were evolved within the framework of local technical possibilities. They enlivened the soil, strengthened the natural resources diversify and maintained the production levels in accordance with the carrying capacity of agro-ecosystem without damaging it.
- ✰ Ancient farmers mostly relied on crop residues, manures, legumes and neem for enriching soil fertility.
- ✰ In Kirishi - parashara, it is stated that crops grown without manure will not give yield and stressed the importance of manures. He also recommended compost preparation from cow dung. The dried, powdered cow dung is placed in pit for decomposition where weed seeds are destroyed. The time duration for composting is two weeks.
- ✰ The use of cowdung, animal bones, fishes, milk as manure are mentioned by Kautilya. Surapala describes the ancient practice of preparing liquid manure prepared by boiling a mixture of animal excreta, bone marrow, flesh, dead fish in an iron pot and then add it to sesame oil cake, honey and ghee.
- ✰ **Preparation of Liquid manure (kunapa):** involves boiling flesh, fat, and marrow 46 Introductory Agriculture of animals such as pig, fish, sheep or goats in water, placing it in earthen pot, and adding milk, powders of sesame oil cake, black gram boiled in honey, decoction of pulses, ghee and hot water. There is no fixed proportion of ingredients. The pot is put in a warm place for two weeks. This fermented liquid manure is called kunapa.
- ✰ **Green manures:** The value of green manure was recognized by farmers in India for thousands of years, as mentioned in treatises like Vrikshayurveda. In Ancient Greece too, farmers ploughed broad bean plants into the soil. Chinese agricultural texts dating back hundreds of

years refer to the importance of grasses and weeds in providing nutrients for farm soil. It was also known to early North American colonists arriving from Europe. Common colonial green manure crops were rye, buckwheat and oats. Traditionally, the incorporation of green manure into the soil is known as the fallow cycle of crop rotation, which was used to allow the soil to regain its fertility after the harvest.

- **In Rajasthan:** *Prosopis cineraria* - brings up moisture and nutrients from the underground and leaves used as green manure.
- **In Tamil Nadu:** *Calotropis gigantiea, Mortinda tinctoria Theprosia purpurea*, Jatropha, Ipomoea Adathoda.
- **In North India :** A traditional weed Kochia indica used as green manure.

☆ Ancient farmers adopted crop rotation and inter cropping to restore soil fertility. Mixed or inter cropping with legumes in cereal and oil seed cultivation were widely practices. All these practices adopted in ancient time are now being recommended today under organic farming concept.

Chapter 6

Irrigation and Water Harvesting

Introduction: An Overview

- ✰ The greatest development in the history of mankind was the discovery of agriculture and then came irrigation. Irrigation has been practiced from very ancient times in our country. Mention of irrigation works in the vedas, puranas and other epics.
- ✰ Often people farm in a place where enough rainfalls during the year (and at the right times) to water the plants just with rainfall. Farmers don't have to worry about the plants getting enough rain. That's called "dry farming" because the farmers don't have to carry water to the plants. But in other places, like Egypt or the Arabian peninsula, it hardly rains at all. Farmers can't rely just on the rainfall to water their crops. They have to find some way of getting water from the river to their fields. That's called "wet farming," and the way they get water from the river is called "irrigation."
- ✰ There are lots of different ways to get water from the river to the fields. One way is just to carry it yourself in buckets, and plenty of people irrigated this way in the ancient and medieval periods (and plenty of people still do today). But it's very hard work carrying water in buckets, and you can't carry very much even if you work very hard. Even a donkey or an ox can't carry enough water to irrigate big fields. So whenever they can, people use some kind of machine to help them carry the water. Often this is a lever, a long wooden pole with a bucket on one end so people can hold the other end of the pole and lower the bucket into the water and then raise it and swing it around and dump it into a canal that is little higher

up and that carried the water to the fields.

Necessity of Irrigation

- The necessity and importance of irrigation for the harvest in India can hardly be exaggerated. It has the potentiality of bringing commercial revolution in any country. British administration in India was also in perfect agreement with this view. The East India Company (EIC) servant Sir Charles Trevelyan noted in 1816 – "that irrigation is everything in India; water is more valuable than land, because when water is applied to land it increases its productiveness at least six-fold and renders great extents of land productive, which otherwise would produce nothing or next to nothing. Despite this, the British administration played a different ball game and the notorious neglect of irrigation in the era of the Company rule proved disastrous as time passed on.
- Marx, in 1853, held that the British rule in India "have neglected entirely" irrigation works and held "the deterioration of an agriculture which is not capable of being conducted on the British principle of free competition, of laissez- faire and lassiez-aller".

History

Ancient India

- The earliest mentions of irrigation are found in Rigveda chapters 1.55, 1.85, 1.105, 7.9, 8.69 and 10.101. The Veda mentions only well-style irrigation, where kupa and avata wells once dug are stated to be always full of water, from which varatra (rope strap) and cakra (wheel) pull kosa (pails) of water. This water was, state the Vedas, led into surmi susira (broad channels) and from there into khanitrima (diverting channels) into fields.
- Later, the 4th-century BCE Indian scholar PâGini, mentions tapping several rivers for irrigation. The mentioned rivers include Sindhu, Suvastu, Varnu, Sarayu, Vipas and Chandrabhaga Buddhist texts from the 3rd century BCE also mention irrigation of crops. Texts from the Maurya Empire era (3rd century BCE) mention that the state raised revenue from charging farmers for irrigation services from rivers.
- Patanjali, in Yogasutra of about the 4th century CE, explains a technique of yoga by comparing it to "the way a farmer diverts a stream from an irrigation canal for irrigation". In Tamil Nadu, the Grand Anicut (canal) across the Kaveri river was implemented in the 3rd century CE, and the basic design is still used today.

Islamic Era

- Waterworks were undertaken during the Delhi Sultanate and the Mughal Empire era from the 12th to 18th centuries. However, these were primarily

to supply water to the palaces and parks of the sultans and other officials.

Colonial Era

- In 1800, some 800,000 hectares was irrigated in India. The British Raj by 1940 built significant number of canals and irrigation systems in Uttar Pradesh, Bihar, Punjab, Assam and Orissa. The Ganges Canal reached 350 miles from Haridwar to Kanpur in Uttar Pradesh. In Assam, a jungle in 1840, by 1900 had 4,000,000 acres under cultivation, especially in tea plantations. In all, the amount of irrigated land multiplied by a factor of eight. Historian David Gilmour states British colonial government had built irrigation network with Ganges canal and that, "by the end of the century the new network of canals in the Punjab" were in place.
- Much of the increase in irrigation during British colonial era was targeted at dedicated poppy and opium farms in India, for exports to China. Poppy cultivation by the British Raj required reliable, dedicated irrigation system. Large portions of the eastern and northern regions of India, namely United Provinces, Northwestern Provinces, Oudh, Behar, Bengal and Rewa were irrigated to ensure reliable supply of poppy and opium for China. By 1850, the Asian opium trade created nearly 1,000 square kilometers of poppy farms in India in its fertile Ganges plains, which increased to over 500,000 acres by 1900. This diversion of food crop land to cash crop use, state scholars, led to massive famines over the 1850 to 1905 period.
- Major irrigation canals were built after millions of people died each in a series of major famines in the 19th century in British India. In 1900, British India (including Bangladesh and Pakistan) had about 13 million ha under irrigation. In 1901 the Viceroy, Lord Curzon, appointed a Commission chaired by Sir Colin Scott-Moncrieff to draw up a comprehensive irrigation plan for India. In 1903 the Commission's report recommended irrigation of an additional 2.6 million hectares. By 1947, the irrigated area had increased to about 22 million ha. In Northwestern British India region alone, with the colonial government's effort, 2.2 million hectares of previously barren land was irrigated by 1940s, most of which is now part of Pakistan. Arthur Cotton led some irrigation canal projects in the Deccan peninsula, and landmarks are named after him in Andhra Pradesh and Tamil Nadu. However, much of the added irrigation capacity during the colonial era was provided by groundwater wells and tanks, operated manually.

Irrigation Trends Since 1947

- India's irrigation covered crop area was about 22.6 million hectares in 1951, and it increased to a potential of 90 mha at the end of 1995, inclusive of canals and groundwater wells. However, the potential irrigation relies of reliable supply of electricity for water pumps and maintenance, and the net irrigated land has been considerably short. According to 2001/2002 Agriculture census, only 58.1 million hectares of land was actually irrigated in India. The total arable land in India is 160 million hectares (395 million acres). According to the World Bank, only about 35 per cent of total agricultural land in India was reliably irrigated in 2010.

Table 6.1: History of Irrigation Development in India

Ancient period	2500-1000 BC	People settled near the banks river/tanks for the purpose getting water for drinking irrigation
Chalcolithic	3000-1700 BC	Practice of irrigation to crops was evolved
Vedic period	1500-1600 BC	People employed craftsman to dig channels from nvers to their fields Well irrigation through kuccha and puccha wells and were practiced
Pandyas/Chola Chera's period	(1st Century 300 AD)	Irrigated rice cultivation started during his period Dams and Tanks were constructed for irrigation
Medieval period	(1200-1700 AD)	Irrigated agriculture was developed during Mogul period Canals, Dams and Tanks were constructed (*e.g.*) 1. Construction of Western Yamuna Canal 2. Constructions of Anantaraja Sagar.

- The ultimate sustainable irrigation potential of India has been estimated in a 1991 United Nations' FAO report to be 139.5 million hectares, comprising 58.5 mha from major and medium river-fed irrigation canal schemes, 15 mha from minor irrigation canal schemes, and 66 mha from groundwater well fed irrigation.
- India's irrigation is mostly groundwater well based. At 39 million hectares (67 per cent of its total irrigation), India has the world's largest groundwater well equipped irrigation system (China with 19 mha is second, USA with 17 mha is third).

Major Irrigation Projects in India

Irrigation Project	*River*
Bhakhra Nangal Project	*Sutlej*
Beas Project	*Beas*
Chambal project	*Chambal*
Damodar Valley project	*Damodar*
Gandak project	*Gandak*
Hirakund project	*Mahanadi*
Indira Gandhi Canal	*Sutlej and Beas*
Kosi Project	*Tapti*
Kalkrapar project	*Tapti*
Kosi Project	*Kosi*
Koyna project	*Koyna*
Malprabha Project	*Malprabha*
Mayurkashi Project	*Mayurkashi*
Nagarjuna Project	*Krishna*
Tungbhadra Project	*Tungbhadra*

History of Irrigation Development in World

- Archaeological investigation has identified evidence of irrigation in Mesopotamia and Egypt as far back as the 6th millennium BCE, where barley was grown in areas where the natural rainfall was insufficient to support such a crop.

Lifting Water from Irrigation Channel using a Swing Basket

Egyptian Irrigation System

Mesopotamia Irrigation System

Pulling Water Out of a Well with Bullock

Kui (Beri)

Bamboo Drip Irrigation System

Figure 6.1a: Some Ancient and Traditional Irrigation System.

Jhalara

Irrigation System at Cantalloc, Nazca, Peru

Figure 6.1b: Some Ancient and Traditional Irrigation System.

- In the 'Zana' Valley of the Andes Mountains in Peru, archaeologists found remains of three irrigation canals radiocarbon dated from the 4th millennium BCE, the 3rd millennium BCE and the 9th century CE. These canals are the earliest record of irrigation in the New World.
- The Indus Valley Civilization in Pakistan and North India (from 2600 BCE) also had an early canal irrigation system. Large scale agriculture was practiced and an extensive network of canals was used for the purpose of irrigation. Sophisticated irrigation and storage systems were developed, including the reservoirs built at Girnar in 3000 BCE.
- There is evidence of the ancient Egyptian pharaoh Amenemhet - III in the twelfth dynasty (about 1800 BCE) using the natural lake of the Fayûm as a reservoir to store surpluses of water for use during the dry seasons, as the lake swelled annually as caused by the annual flooding of the Nile.
- The Qanats, developed in ancient Persia in about 800 BCE, are among the oldest known irrigation methods still in use today. They are now found in Asia, the Middle East and North Africa. The system comprises a network of vertical wells and gently sloping tunnels driven into the sides of cliffs and steep hills to tap groundwater.
- The Noria, a water wheel with clay pots around the rim powered by the flow of the stream (or by animals where the water source was still), was first brought into use at about this time, by Roman settlers in North Africa. By 150 BCE pots were fitted with valves to allow smoother filling as they were forced into the water.
- The irrigation works of ancient Sri Lanka, the earliest dating from about 300 BCE, in the reign of King Pandukabhaya and under continuous development for the next thousand years, were one of the most complex

irrigation systems of the ancient world. In addition to underground canals, the Sinhalese were the first to build completely artificial reservoirs to store water. The system was extensively restored and further extended during the reign of King Parakrama Bahu (1153 – 1186 CE). In the Szechwan region ancient China the Dujiangyan Irrigation System was built in 250 BCE to irrigate a large area and it still supplies water today.

- In fifteenth century Korea the world's first water gauge (woo ryang gyae) was discovered in 1441 CE. The inventor was Jang Young Sil, a Korean engineer of the Choson Dynasty, under the active direction of the King, Se Jong. It was installed in irrigation tanks as part of a nationwide system to measure and collect rainfall for agricultural applications. With this instrument, planners and farmers could make better use of the information gathered in the survey.

Present Status of Irrigation Development in India and World

- At the global scale 278.8 Mha (689 million acres) of agricultural land was equipped with irrigation infrastructure around the year 2000. About 68 per cent of the area equipped for irrigation is located in Asia, 17 per cent in America, 9 per cent in Europe, 5 per cent in Africa and 1 per cent in Oceania. India's irrigation development in this century, and particularly after independence, has seen large number of large storage based systems, all by the government effort and money.
- Post independence has seen more than 60 per cent of irrigation budgets going for Major and Medium (M and M) projects. However, Growth rate of irrigated area continues to fall from 4.23 per cent per year during the 1970s to 3.08 per cent per year in 1980s and to 2.56 per cent in the 1990s. At present, with almost one fifth of worlds net irrigated area (57 Mha); India has the highest irrigated area in the world today. India's ultimate irrigation potential was estimated at 139.9 Mha, comprising of 58.46 Mha through major and medium irrigation schemes and 81.43 Mha from minor irrigation schemes.
- Recently some positive steps were also taken to long-awaited inter-basin water transfer, aiming at adding 35 Mha to India's irrigated area. The implementation of the inter-basin water transfer link schemes are taken up in a phased manner depending on the priorities of the Government. The links namely (i) Ken-Betwa link (ii)Parbati-Kalisindh-Chambal link (iii) Godavari (Polavaram) – Krishna(Vijayawada) link (iv) Damanganga-Pinjal link and (v) Par-Tapi-Narmada Link have been identified as priority links for consensus building amongst concerned States for taking up preparation of Detailed Project Reports (DPR).

History of Rainwater Harvesting

An Overview

- ✰ The capturing and storing of rainwater goes back thousands of years to when we first started to farm the land and needed to find new ways of irrigating crops. In hotter climes, catching that intermittent rainfall often meant the difference between life and death for communities. Whilst the need to conserve water fell away with greater urbanisation in the last thousand years, we are once again returning to this ancient and vital part of greener living.
- ✰ The Rainwater harvesting is the simple collection or storing of water through scientific techniques from the areas where the rain falls. It involves utilization of rain water for the domestic or the agricultural purpose. The method of rain water harvesting has been into practice since ancient times. It is as far the best possible way to conserve water and awaken the society towards the importance of water. The method is simple and cost effective too. It is especially beneficial in the areas, which faces the scarcity of water.

People usually make complaints about the lack of water. During the monsoons lots of water goes waste into the gutters. And this is when Rain water Harvesting proves to be the most effective way to conserve water. We can collect the rain water into the tanks and prevent it from flowing into drains and being wasted. It is practiced on the large scale in the metropolitan cities. Rain water harvesting comprises of storage of water and water recharging through the technical process.

Rainwater Harvesting in Ancient Times

- ✰ Civilisations in the Indus Valley were far more advanced than we may think nowadays. In many of the ancient cities that still remain, we can still find huge vats that were cut into the rock to collect water when there was torrential rainfall. These were used to keep the population and local vegetation going in hotter, dryer times and were fed by numerous stone gullies that weaved their way through the city. Some of these rock vats are still used today in parts of India.
- ✰ Another technique that has been used for hundreds of years in India is to build water harvesting systems on top of the roofs of houses. It's a simple technology that has spread across the world, particularly to countries such as Brazil and China.
- ✰ The technology of rainwater harvesting is deeply rooted in the social fabric of India with a variety of ancient methods still found today. These include:

- *Talibs:* Medium to large sized reservoirs that provide irrigation for plants as well as drinking.
- *Johads:* Dams that are used to capture and keep rainwater.
- *Baoris:* Wells dug into the ground that are often still used for drinking.
- *Jhalaras:* Specially constructed tanks that are used for the local community and religious purposes.

The Romans and Rainwater Harvesting

- During the time of the Roman Empire, rainwater collection became something of an art and science, with many new cities incorporating state of the art technology for the time. The Romans were masters at these new developments and great progress was made right up until the 6th Century AD and the rule of Emperor Caesar.
- One of the most impressive rainwater harvesting constructions can be found in Istanbul in the Sunken Palace which was used to collect rainwater from the streets above. It's so large that you can sail around it in a boat.

Later Rainwater Harvesting

- In the 17th Century the small island of Malta built an aqueduct to collect rainwater for its growing population. It was a popular way of getting water to the people but as new methods of building houses and supplying water improved with things like water pipes and reservoirs, the technology of rainwater harvesting stalled over the following centuries.
- The other thing that stopped using collected rainwater for many towns and cities was also the prospect of spreading disease. The Sunken Palace in Istanbul stopped using the water for drinking many hundreds of years ago and the caches that Arab tribes kept across desert lands have mainly fallen into disuse. With increased urbanisation, the need for effective rainwater catchment hasn't been a major concern except in countries where the climate dictates it and water is in short supply.

Some Traditional Methods of Conserving Rain Water

- In ancient days, especially Indians, know the methods of conservation of rain water. There are evidences that, even during Harappan period, there was very good system of water management as could be seen in the latest excavation at Dholavira in Kachch. During Independence period, the people use to manage water resources considering it as part of the nature which is essential for their survival. This could be seen from the rain water harvesting structures in the low rainfall areas of Rajasthan, harvesting springs in hilly areas and mountainous region and percolation ponds and tanks in southern India.

- The ancient people in Tamil Nadu stored rainwater in public, placed separately one for drinking purposes and another for bathing and other domestic purposses and called them as Ooranies. They also formed percolation tanks or ponds, for the purpose of recharging irrigation or domestic wells. They periodically clean the water ways so as to get clean water throughout the year. These are instances in the history that people constructed crude rubble bunds across river courses either for diversion of water or for augmenting the ground water.

Bamboo Method of Rainwater Harvesting

- In Meghalaya (one of the seven northeastern states in India), an indegenious system of tapping of stream and springwater by using bamboo pipes to irrigate plantations is widely prevalent. It is so perfected that about 18-20 litres of water entering the bamboo pipe system per minute gets transported over several hundred metres and finally gets reduced to 20-80 drops per minute at the site of the plant. The tribal farmers of Khasi and Jaintia hills use the 200-year-old system.
- The bamboo drip irrigation system is normally used to irrigate the betel leaf or black pepper crops planted in arecanut orchards or in mixed orchards. Bamboo pipes are used to divert perennial springs on the hilltops to the lower reaches by gravity. The channel sections, made of bamboo, divert and convey water to the plot site where it is distributed without leakage into branches, again made and laid out with different forms of bamboo pipes. Manipulating the intake pipe positions also controls the flow of water into the lateral pipes. Reduced channel sections and diversion units are used at the last stage of water application. The last channel section enables the water to be dropped near the roots of the plant.

Kunds of Thar Desert

- In the sandier tracts, the villagers of the Thar Desert had evolved an indegenious system of rainwater harvesting known as kunds or kundis. Kund, the local name given to a covered underground tank, was developed primarily for tackling drinking water problems. Usually constructed with local materials or cement, kunds were more prevalent in the western arid regions of Rajasthan, and in areas where the limited groundwater available is moderate to highly saline.
- Groundwater in Barmer, for instance, in nearly 76 per cent of the district's area, has total dissolved salts (TDS) ranging from 1,500-10,000 parts per million (ppm). Under such conditions, kunds provide convenient, clean and sweetwater for drinking. Kunds were owned by communities or privately, with the rich having one or more kunds of their own. Community

kunds were built through village cooperation or by a rich man for the entire community.

Traditional Rainwater Harvesting

- The traditional rainwater harvesting methods in North India is surface water harvesting methods are *viz.,* Tanka, Nada, Nadi, Talai, Talab, Khadin Sar, Sagar and Samend. Depending upon rainfall, topography of area, type of soil, the water harvesting methods are different from region to region.
  - *Tanka:* It is one of the ancient, common and relatively hygenic methods of water storage. It is constructed of on farm, country yard and fort. The shape is normally circular/square. Dimension is 2 m dia. 3 m deep capacity 10000 lit. It is made on sloping land to arrest run off water in the farm; however, in houses the construction is made on an elevated place to avoid entry of dirty water in to it.
  - *Talai:* Similar to Tanka, still deeper (2-3cm depth). Special attention paid for selection of location such that there is adequate flow of rain water into Talai. Care is also taken so that loose soil does not flow along with water stream.
  - *Nada:* In this method, low lying areas in between hillocks is excavated as pit and provided embankment to arrest rain water from these hillocks. The catchment area of Nada is 5 to 10 ha. The Nada is constructed on rangeland, barren land pastureland and agriculture field. It provides short-term storage of rainwater and mainly used for animals.
  - *Nadi:* Compared to Nada, the Nadi is bigger in size. A village or group of Villages uses the runoff water collected in the Nadi. Depth is 6-8m, catchment area 10-150 ha. In the Nadi, water is available for whole of the year as a result it provides shelter for many wild animals and birds.
  - *Talab:* It is relatively shallow and spread over to more area compared to Nadi. It is generally constructed in rangeland. The catchment area of Talab is 480 ha, when it is filled its fullest capacity can lost for many years.
  - *Khadin:* Accumulation of runoff water in between hillocks is known as Khadin. Khadin means cultivation crops. The khadin water is generally used for crop cultivation and animals.
  - *Sar, Sagar and Samand:* It is used to harvest rainwater for irrigation purpose. Even today this structure provides excellent source of reservoir and also tourist spot. Practices of irrigation and rainwater harvesting adopted in ancient period were more relevant in Indian agriculture today.

Modern Rainwater Harvesting

- Go into any garden store in the UK and you will no doubt be able to see various plastic butts that are designed to collect rainwater so that we can water our gardens and keep the plants in good health during any dry period. There's no doubt that climate change has got us thinking about water conservation again, especially with the supply companies beginning to put their prices up. The general consensus is that letting all that rainwater go to waste is no longer acceptable. It can be collected and that can help reduce water bills. In other words, is not only a good idea ecologically, it makes sense financially.
- Recycled water can be used for a variety of daily tasks from washing clothes, flushing toilets and even cleaning the car. With the possibility that we could face more prolonged drier periods in the near future, the onus is on us all to conserve what we consume and make the most of what comes to us free of charge from the sky.

Methods of Rainwater Harvesting

- *Catchment:* Any surface or the paved areas can be treated as catchment. Even the footpaths and roads can act as the catchment, as these areas too receive the direct rainfall. Rooftops are the best among them because of the large coefficient of run-off generated from them and there are less chances of contamination of water.
- *Conveyance:* Conveyance system basically includes rain gutters and down pipes which collects the rain water from catchment to the storage tank. These rain gutters are usually built during the time of construction. They need to be designed appropriately as to avoid the loss of water during the conveyance process.
- *Storage:* The most important part of the rain water harvesting is the storage system. The storage system is designed according to the amount of water that is to be stored. The design and site (location) of the storage or the recharge system should be properly chosen. The areas which receives the rainfall frequently, there a simple storage system could be constructed, to meet the daily water requirements. Otherwise the areas which receive the lesser rainfall, there the storage systems are quite essential. Rain barrels, underground or open slumps are mostly used to collect rain water. Make sure that the storage system is properly sealed and does nor leak. Use Chlorine from time to time to keep the water clean.

Benefits of Rainwater Harvesting

- Rainwater harvesting first of all increases water security. It is the perfect solution to meet water requirements especially in the areas which do not have sufficient water resources.

- It helps in improving the quality of the ground water and increasing the level of the ground level. It also helps in improving the overall floral system.
- It reduces the loss of top layer of the soil. If we capture the water directly, we need not to depend much on the water storage dams.
- It is the good solution to the increasing water crises.
- Rain water harvesting reduces the flooding on roads and further prevents it from contamination. And in the last it decreases the menace of floods on regional scale.

Chapter 7

Description of Indian Civilization and Agriculture by Travellers

An Introduction

The tradition of Chinese, European and Americans travelling to India dates back to the days of the ancient Greek era. The middle ages saw Europeans, Chinese and Americans from different nations exploring India with both agricultural and commercial motivations.

The views of the Chinese, European and Americans travellers are described as under:

Indus Valley Civilization

- ✰ The types of crops that the Indus Civilization had was wheat, barley, peas, lentils, linseed, and mustard. Experts say that they might have grown cotton in the summer. They did not grow well where they lived, but they did find white rice and fed it to their animals. The silt that the river brought in when it flooded was the reason why they can grow this many crops. The nutrients that the plants needed was replenished every year when the annual floods came in.
- ✰ Indus valley civilization: Allchins, relying on Lambrick, who, according to them, had personal knowledge of Sind, describe as follows how crops were grown in the riverain tract of the Indus.
- ✰ "The principal food grains, that is wheat and barley, would have been grown as spring (*rabi*) crops: that is to say, sown at the end of the

inundation upon land which had been submerged by spill from the river or one of its natural flood channels, and reaped in March or April.

Megasthenes (born BC.350 BC-died C.290)

- ☆ The Greek writers highly praised the fertility of Indian soil and favourable climate condition describing the principal agricultural products of the land.
- ☆ The Greek writers also affirm that India has a double rainfall and the Indians generally gather two harvests. - Megasthenes witnesses - the sowing of wheat in early, winter rains and of rice, 'bosporum', sesamum and millets in the summer solstice (Diodorus, II, 36).
- ☆ The Greek historian and diplomat Megasthenes adds further to the winter crops, *viz.*, wheat, barley, pulse and other esculent fruits unknown to us.

Hiuen Tsang's

- ☆ The Chinese traveller Hiuen Tsang visited India during the period of emperor Harsha. When he went back to China, he wrote a detailed description of India during the reign of Harsha in his book 'Si-yu-ki'or 'Record of the Western Countries'.
- ☆ The primary aim of the visit of Hiuen Tsang to India was to gain knowledge of Buddhism and collect its religious texts. As he did not get the permission of the Chinese emperor to visit India, he slipped away from there in 629 A.D. He crossed the desert of Gobi, visited several places in Central Asia like Kashagar, Samarkand and Balkha and reached Afghanistan. He met and found worshippers of the Sun, a large number of Buddhist monks and followers, Stupas and monasteries at different places.
- ☆ From Afghanistan he reached Taxila via Peshawar. The journey from China to India was covered by him in about a year. Then he stayed in India for nearly fourteen years. From Taxila, he went to Kashmir and then visited several places in India like Mathura, Kannauj, Sravasti, Ayodhya, Kapilvastu, Kusinagara, Sarnath, Vaisali, Pataliputra, Rajagraha, Bodha-Gaya and Nalanda.
- ☆ He remained at the University of Nalanda for about five years. He, then, proceeded to Bengal and visited South India as well, as far as Kanchi. He had been a guest to Bhaskara Varman, ruler of Kamarupa. From there he was called to the court of Harsha. Harsha called a religious assembly at Kannauj to honour him. Hiuen Tsang presided over that assembly.
- ☆ He also participated in one of the religious assemblies called by Harsha at Prayag after that. He left India in 644 A.D. through the same route by which he had entered. He took back many images of Buddha and copies of different Buddhist religious texts. When he reached back China he

was received with honour by the Chinese emperor. Then he wrote the description of India at the instance of the Emperor.

- Hiuen Tsang described the city-life of India. The information we gather from his account is that the houses were of varied types and were constructed with wood, bricks and dung. The city-streets were circular and dirty. Many old cities were in ruins while new cities had grown up.
- He described that Indians used cotton, silk and wool for their garments and these were of varied types. He described Indians as lovers of education, literature and fine arts.
- Hiuen Tsang described the social condition of India in detail. He wrote that caste-system was rigid. There was no purdah-system and women were provided education. However, the practice of sati prevailed. In general, the common people were simple and honest.
- They used simple garments and avoided meat, onions and liquor in their food and drinks. They observed high morality in their social and personal lives. The rich people dressed well, lived in comfortable houses and enjoyed all comforts and amenities of life.
- Hiuen Tsang also wrote about the economic condition of India at that time. He gave a long list of Indian fruits and agricultural products. India produced the best cotton, silk and woolen cloth at that time and prepared all sorts of garments from them. He praised very much the quality of Indian pearls and ivory.

Protection of Cultivators

- *Sher Shah Suri* (1486 – 22 May 1545), had genuine concern for the peasantry and safety of their crops. One of the regulations made by Sher shah was this: That his victorious standards should cause no injury to the cultivations of the people; and when he marched he personally examined into the state of the cultivation, and stationed horsemen round it to prevent people from trespassing on any one's field.
- As regards the peasantry and their condition, there is reliable evidence in the observations of the European travellers who travelled in India in the seventeenth century.
- Evidence of the structure of the Mughal gardens and plants grown in them is in the Persian classics illustrated during the reign of Akbar. Among them is *Diwan-i-Anwari,* a collection of poems by the Persian poet Anwari, who flourished in the latter part of twelfth century. It contains some excellent paintings on gardens and gardening.
- *Abu-l-Fazl* describes three kinds of sugarcane, *viz.* paunda, black and ordinary and he also provides a list of twenty-one fragrant flowering plants along with the colour of their flowers and the season of flowering in the Ain- iAkbari.

- Sugarcane was the most widely grown cash crops. The Ain-I-Akbari records it in most of the dastur cricles of Agra, Awadh,Lahore, Multan and Allahabad. Sugar from Bengal was considered to be the best in quality. Multan, Malwa, Sind, Khandesh, Berar and regions of South India all testify to the presence of the sugarcane in the 17th century. Cultivation of tobacco seems to have spread in India in a short time. The Ain-I-Akbari does not mention it as a crop in any of the dastur circles or other regions. It seems to have been introduced to India during the 16th century by the Portuguese. Its cultivation was noticed in almost all parts of he country (specially Surat and Bihar).

Terry

- An English traveler, writes, 'The country was abounding with musk-melons. One could also find water-melons, pomegranates, lemons, oranges, dates, figs, grapes, coconut, plantains, mangoes, pineapples, pears, apples, *etc*.'
- Terry also mentions the use of coffee by some people and he writes, 'Many religious people drank a "wholesome liquor" which they called coffee. Black seeds were boiled in water, which also become black. It altered the taste of water very little. It quickened the spirit and cleansed the blood.

Francois Bernier

- Francois Bernier (1620 - 1688) was born in Joue, near Gonnord, in Anjou to a family of cultivators of the soil. Concurrent with his birth, was Shah Jahan's accession to the Mughal throne in 1628.
- Towards the end of the 18th century, the volume of French accounts of India grew substantially and a peculiar tendency emerged. The late 18th century French images of Hindustan or 'Bharat' were painted with glorifying shades of Indomania. Their astonishment at the sheer size of the Indian continent, as they called it, was surpassed only by their praise for Indian political organization and agriculture. They envisioned India as a continent that was the progenitor of religious and scientific wisdom and was metamorphosing through their renaissance by the sheer spirit of their religious fervour for agriculture (Wink, 450).
- *Francois Bernier:* Of the European travelers who come to India during the Mughal rule, the most intelligent and learned was Francois Bernier a Frenchman. He gives a vivid description of Bengal its landscape people and its plant and animals products. With extensive fields of rice, sugar, corn, three or four sorts of vegetables, mustard, sesames for oils and small mulberry trees two or three feet (61 to 91 cm) in height, for the food of silk worms.

Coloner Philip Meadows Taylor

- Meadows Taylor (1808-1876) states "The Bahmanis constructed irrigation works in the eastern provinces, which incidentally did good to the peasantry while primarily securing the crown revenue.
- He made a number of contributions the Gulburga region in India, by initiating a number of reforms. He encouraged improvement of Agriculture, opened up more job opportunities, started schools and improve infrastructure. He was known to spend his own money to provide a drought relief.

Herodotus

- Herodotus (484-425 BC) the father of history reported in his writings that the wild Indian (cotton) trees possessed in their fruits fleeces, superseding those of sheep in beauty and excellence from which the natives used to weave cloth. Herodotus further wrote that "trees which grow wild in India and the fruit of which bear wool exceeding in beauty and fineness that of sheep wool Indians make their clothes with this tree wool". Some traveller writers fabricated stories of a lamb sitting inside the fruit.
- The Herodotus, wrote in 400 BC that in India there were "trees growing wild, which produce a kind of wool better than sheep's wool in beauty and quality, which the Indians use for making their clothes." During this period, the famous Ajanta Cave carvings show innovating cotton growers in India had invented an early roller machine to get the seeds out of the cotton.

Marco Pola

- **Marco Pola,** a Venetian, who travelled widely throughout the Asia in AD 1290 said that the coast of Coromandel (Madras, India) produced the finest and most beautiful cotton in the world. Indian cloth, particularly the Dacca muslin was renowned all over the world and has been described as 'webs of woven wind' by oriental poets. It was so fine that it could hardly be felt in the hands. It is said that when such muslins were laid on the grass to bleach and the dew had fallen, it was no longer visible.
- A whole garment made from it could be drawn through a wedding ring of medium size. There is also the often repeated tale of Mughul princes who put on seven layers of muslin and still the contours of her body were so visible that she had to be admonisher by her father, Muhammed Bin Thuglak.
- By the Gupta period, circa 200 AD, the Indians were selling cotton as a luxury good to their neighbours in the east and west – the Chinese and the Parthians. Further west, the Roman considered cotton as luxurious and

as expensive as silk, which they bought from Arabic or Parthian traders. Like Herodotus, the Roman author and philosopher Pliny wrote that in India there were, "trees that bear wool" and "balls of down from which an expensive linen material for clothes is made.

Chapter 8

Indian Agriculture from Past to Modern Era

Introduction: An Overview

> *Food drives the world; apart from clean water, access to adequate food is the primary concern for most people on earth. This makes agriculture one of the largest and most significant industries in the world; agricultural productivity is important not only for a country's balance of trade but the security and health of its population as well.*

- ☆ The rapid change in Indian Agriculture particularly since the era of green revolutio n has been received with mixed response. The development perspective has been definitely encouraging and viewed positively by almost all sectors. However, at the same time, the conservative view has echoed the need for the system sustainability, which may also be arguably right. Both viewpoints have complementarity and also distinct application values, which need to be accepted. Therefore, one should always look forward to addressing both sustainability and development in Indian agriculture in the years to come. The changing scenario has recorded agriculture more as a business proposition than merely tradition at least in the agriculturally progressive states. The traditional agriculture, which was a way of life of the majority of Indian farmers, has gradually shrunken to fewer pockets.
- ☆ It is rightly believed that the traditional farming was consistent for cropping patterns, mixed crop composition, cultivation practices, *etc.*, and it was based on the hereditary and the community level experiences. This way of agriculture harboured several friendly and synergistic biota,

including, earthworms and soil micro-organisms. The non-biological resources were equally nurtured, for example, the soil structure, composition and organic matter content. In terms of subsistence, the traditional agriculture ensured the farming families more or less self-sufficiency because of an array of harvests even though individual commodities harvested were little in quantities. The farmer did not have to pay for these commodities in the local market. Rather they sometimes earned a little profit by selling their marginal surplus in the weekly haats.

- The major agricultural products can be broadly grouped into foods, fibers, fuels, and raw materials. In the 21st century, plants have been used to grow biofuels, biopharmaceuticals, bioplastics, and pharmaceuticals. Specific foods include cereals, vegetables, fruits and meat. Fibres include cotton, wool, hemp, silk and flax. Raw materials include lumber and bamboo.
- Other useful materials are produced by plants. Such as resins. Biofuels include methane from biomass, ethanol, and biodiesel. Cut flowers, nursery plants, tropical fish and birds for the pet trade are some of the ornamental products.

Ancient Origins

- The Fertile Crescent of Western Asia, Egypt, and India were sites of the earliest planet sowing and harvesting of plants that had previously been gathered in the wild. Independent development of agriculture occurred in northern and southern China, Africa's Sahel, New Guinea and several regions of the Americas. The eight so-called Neolithic founder crops of agriculture appear: first emmer wheat and einkorn wheat, then hulled barley, peas, lentils, bitter vetch, chick peas and flax.

The potato, tomato, pepper, squash, several varieties of bean, tobacco, and several other plants were also developed in the New World, as was extensive terracing of steep hillsides in much of Andean South America. The Greeks and romans built on techniques pioneered by the Sumerians, but made few fundamentally new advances. Southern Greeks struggled with very poor soils, yet managed to become a dominant society for years. The Romans were noted for an emphasis on the cultivation of crops for trade.

- By 7000BC, small-scale agriculture reached Egypt. From at least 7000BC the Indian subcontinent saw farming of wheat and barley, as attested by archaeological excavation at Mehrgarh in Balochistan in what is present day Pakistan. By 6000BC, mid-scale farming was entrenched on the bank of the Nile. About this time, agriculture was developed independely in the Far East, with rice, rather than wheat, as the primary crop. Chinese and Indonesian farmers went on to domesticate taro and beans including mung, soy and azuki. To complement these new sources of carbohydrates, highly organized net fishing of river, lakes and ocean shores in these areas

brought in great volumes of essential protein. Collectively, these new methods of farming and fishing inaugurated a human population boom that dwarfed all previous expansions and continues today.

- ☆ By 5000BC, the Sumerians had developed core agricultural techniques including large scale intensive cultivation of land, monocropping, organized irrigation, and the use of a specialized labour force, particularly along the waterway now known as the Shatt al-Arab, from its Persian Gulf delta to the confluence of the Tigris and Euphrates. Domestication of wild aurochs and mouflon into cattle and sheep, respectively, ushered int eh large-scale use of animals for food/fibre and as beasts of burden. The shepherd joined the farmer as an essential provider for sedentary and seminomadic societies. Maize, manioc, and arrowroot were first domesticated in the Americas as far back as 5200BC.
- ☆ In the Americas, a parallel agricultural revolution occurred, resulting in some of the most important crops grown today. In Mesoamerica wild teosinte was transformed through human selection into the ancestor of modern maize, more than 6000 years ago. It gradually spread across North America and was the major crop of Native Americans at the time of European exploration. Other Mesoamerican crops include hundreds of varieties of squash and beans. Cocoa was also a major crop in domesticated Mexico and central America. The turkey, one of the most important meat birds, was probably domesticated in Mexico or the U.S. Southwest. In the Andes region of South America the major domesticated crop was potatoes, domesticated perhaps 5000 years ago. Large varieties of beans were domesticated, in South America, as well as animals, including llamas, alpacas, and guinea pigs. Coca, still a major crop, was also domesticated in the Andes.
- ☆ A minor center of domestication, the indigenous people of the Eastern U.S. appear to have domesticated numerous crops. Sunflowers, tobacco, varieties of squash and Chenopodium, as well as crops no longer grown, including marshelder and little barley were domesticated. Other wild foods may have undergone some selective cultivation, including wild rice and maple sugar. The most common varieties of strawberry were domesticated from Eastern North America.

Early History

- ☆ Barley and wheat cultivation - along with the rearing of cattle, sheep and goat - was visible in Mehrgarh by 8000-6000 BCE. Agro pastoralism in India included threshing, planting crops in rows - either of two or of six - and storing grain in granaries. In the period of the Neolithic revolution (roughly 8000-4000 BCE.), agriculture was far from the dominant mode of support for human societies. But those who adopted it, have survived and increased, and passed their techniques of production to the next generation. This transformation of knowledge was the base of further development in agriculture. By the 5th millennium BCE agricultural communities became widespread in Kashmir. Zaheer Baber (1996) writes that 'the first evidence of cultivation of cotton had already developed'. Cotton was cultivated by the 5th millennium BCE-4th millennium BCE. The Indus cotton industry was well developed and some methods used in cotton spinning and fabrication continued to be practiced till the modern Industrialization of India.

- A variety of tropical fruit such as mango and muskmelon are native to the Indian sub-continent. The Indians also domesticated hemp, which they used for a number of applications including making narcotics, fibre, and oil. The farmers of the Indus valley, which thrived in modern-day Pakistan and North India, grew peas, sesame, and dates, sugarcane was originally from tropical South Asia and Southeast Asia. Different species likely originated in different locations with *S. barberi* originating in India and *S. edule* and *S. officinarum* coming from New Guinea. Wild rice cultivation appeared in the Belan and Ganges valley regions of northern India as early as 4530 BCE and 5440 BCE respectively. Rice was cultivated in the Indus Valley Indus valley civilization Agricultural activity during the second millennium BC included rice cultivation in the Kashmir and Harrappan regions. Mixed farming was the basis of the Indus valley economy. Denis J. Murphy (2007) details the spread of cultivated rice from India into South-east Asia.
- Several wild cereals, including rice, grew in the Vindhya Hills, and rice cultivation, at sites such as Chopani-Mando and Mahagara, may have been underway as early as 7000 BP. The relative isolation of this area and the early development of rice farming imply that it was developed indigenously - Chopani-Mando and Mahagara are located on the upper reaches of the Ganges drainage system and it is likely that migrants from this area spread rice farming down the Ganges valley into the fertile plains of Bengal, and beyond into south-east Asia.
- Irrigation was developed in the Indus Valley Civilisation by around 4500 BCE. The size and prosperity of the Indus civilisation grew as a result of this innovation, which eventually led to more planned settlements making use of drainage and sewers. Sophisticated irrigation and water storage systems were developed by the Indus Valley Civilisation, including artificial reservoirs at Girnar dated to 3000 BCE, and an early canal irrigation system from circa 2600 BCE. Archeological evidence of an animal-drawn plough dates back to 2500 BC in the Indus Valley Civilisation.
- Outside of the Indus Valley area of influence there are 2 regions with distinct agricultures dating back to around 2800-1500 BCE. These are the Deccan Plateau and an area within the modern states of Odisha and Bihar.

Within the Deccan the ashmound tradition developed c.2800 BCE. This is characterised by large mounds of burn cattle dung and other materials. The people of the ashmound tradition grew millets and pulses, some of which were domesticated in this part of India, for example, *Brachiaria rasoma, Setaria verticillata, Vigna radiata* and *Macrotyloma uniflorum*. They also herded cattle, sheep and goat and were largely engaged in pastoralism (Fuller 2006, 'Dung mounds and Domesticators'). In the east of India Neolithic people grew rice and pulses, as well as keeping cattle, sheep and goat. By 1500 BCE a distinct agriculture focused on summer crops, including Vigna and *Panicum milliaceum* was developed.

Vedic Period – Post Maha Janapadas Period (1500 BCE – 200 CE)

- Vedic literature provides some of the earliest written record of agriculture in India. Rigveda hymns, for example, describes plowing, fallowing, irrigation, fruit and vegetable cultivation. Other historical evidence suggests rice and cotton were cultivated in the Indus Valley, and plowing patterns from the Bronze Age have been excavated at Kalibangan in Rajasthan, Bhumivargaha, another ancient Indian Sanskrit text, suggested to be 2500 years old, classifies agricultural land into twelve categories: urvara (fertile), ushara (barren), maru (desert), aprahata (fallow), shadvala (grassy), pankikala (muddy), jalaprayah (watery), kachchaha (land contiguous to water), sharkara (full of pebbles and pieces of limestone), sharkaravati (sandy), nadimatruka (land watered from a river), and devamatruka (rainfed). Some archaeologists believe rice was a domesticated crop along the banks of the Indian river ganges in the sixth millennium BC. So were species of winter cereals (barley, oats, and wheat) and legumes (lentil and chickpea) grown in Northwest India before the sixth millennium BC.
- Other crops cultivated in India 3000 to 6000 years ago, include sesame, linseed, safflower, mustards, castor, mung bean, black gram, horse gram, pigeonpea, field pea, grass pea (khesari), fenugreek, cotton, jujube, grapes, dates, jackfruit, mango, mulberry, and black plum. Indian peasants had also domesticated cattle, buffaloes, sheep, goats, pigs and horses thousands of years ago. Some scientists claim agriculture in India was widespread in the Indian peninsula, some 3000–5000 years ago, well beyond the fertile plains of the north. For example, one study reports twelve sites in the southern Indian states of Karnataka and Andhra Pradesh providing clear evidence of agriculture of pulses (*Vigna radiata* and *Macrltyloma uniflorum*), millet-grasses (*Brachiaria ramosa* and *Setaria verticillata*), wheats (*Triticum dicoccum, Triticum durum/ aestivum*), barley (*Hordeum vulgare*), hyacinth bean (*Lablab purpureus*), pearl millet (*Pennisetum glaucum*), finger millet (*Eleusine coracana*),

cotton (*Gossypium* sp.), linseed (*Linum* sp.), as well as gathered fruits of Ziziphus and two Cucurbitaceae.

- ☆ Some claim Indian agriculture began by 9000 BP as a result of early cultivation of plants, and domestication of crops and animals. Settled life soon followed with implements and techniques being developed for agriculture. Double monsoons led to two harvests being reaped in one year. Indian products soon reached the world via existing trading networks and foreign crops were introduced to India. Plants and animals - considered essential to their survival by the Indians - came to be worshiped and venerated. The middle ages saw irrigation channels reach a new level of sophistication in India and Indian crops affecting the economies of other regions of the world under Islamic patronage. Land and water management systems were developed with an aim of providing uniform growth. Despite some stagnation during the later modern era the independent Republic of India was able to develop a comprehensive agricultural programme.
- ☆ It is likely that summer monsoons may have been longer and may have contained moisture in excess than required for normal food production. One effect of this excessive moisture would have been to aid the winter monsoon rainfall required for winter crops. In India, both wheat and barley are held to be Rabi (winter) crops and - like other parts of the world - would have largely depended on winter monsoons before the irrigation became widespread. The growth of the Kharif crops would have probably suffered as a result of excessive moisture. Jute was first cultivated in India, where it was used to make ropes and cordage. Some animals - thought by the Indians as being vital to their survival - came to be worshiped. Trees were also domesticated, worshiped, and venerated - Pipal and Banyan in particular. Others came to be known for their medicinal uses and found mention in the holistic medical system Ayurveda.
- ☆ In the later Vedic texts (c. 3000 -2500 BP) there are repeated references to agricultural technology and practices, including iron implements; the cultivation of cereals, vegetables, and fruits; the use of meat and milk and animal husbandry. Farmers plowed the soil broadcast seeds, and used a certain sequence of cropping and fallowing. Cow dung provided fertilizer, and irrigation was practiced. The Mauryan Empire (322–185 BCE) categorised soils and made meteorological observations for agricultural use. Other Mauryan facilitation included construction and maintenance of dams, and provision of horse-drawn chariots - quicker than traditional bullock carts. The Greek diplomat Megasthenes (c. 300 BC) - in his book *Indika* - provides a secular eyewitness account of Indian agriculture.
- ☆ India has many huge mountains which abound in fruit-trees of every kind, and many vast plains of great fertility. The greater part of the soil, moreover, is under irrigation, and consequently bears two crops in the

course of the year. In addition to cereals, there grows throughout India much millet and much pulse of different sorts, and rice also, and what is called bosporum [Indian millet]. Since there is a double rainfall [*i.e.*, the two monsoons] in the course of each year the inhabitants of India almost always gather in two harvests annually.

Agriculture in India

Agriculture in India, the preeminent sector of the economy, is the source of livelihood of almost two third of the workforce in the country. The contribution of agriculture and allied activities of India's economic growth in recent years has been no less significant than that of industry and services. The importance of agriculture to the country is best summed up by this statement: "If agriculture survives, India survives".

Age-old Indian Farms in Agri Hall of Frame

- Koraput in Odisha and Kuttanad in Kerala are among a crowded list of the Globally Important Agricultural Heritage Systems.
- To be eligible, the farming practice has to have a "current global significance".
- In Kuttanad, large-scale cereal farming takes place below the sea level.
- Koraput has been identified as one of the centres of origin of rice itself.

Two of India's ancients farming system, intact over 2,000 years and still relevant, have been shortlisted by the UN's Food and Agricultural Organization (FAO) as world agriculture heritage site, being compiled for the first time. Koraput in Odisha and Kuttanad in Kerala are among a crowded list of the Globally Important Agricultural Heritage Systems, including south Italy's Lemon Gardens and Iran's Qashqai pastoral nomads. A third farming system, from India's Northeast, is still being evaluated for qualification. To be eligible, it is not enough for a farming practice to be age-old; it has to have a "current global significance". Legally, the sites will have the same status and require similar protection as the UN's other well-known world Heritage Sites, such as the Taj Mahal, Kuttanad, an idyllic farm belt hugging Kerala's backwaters, is the only place where largescale cereal farming takes place below the sea level. Situated in Alappuzha district, it is the rice bowl of Kerala. Koraput has been identified as one of the centres of origin of rice itself, the staple food of more than half the world's population. Koraput's tribals have cultivated 79 different species of cereals, pulses and millets, one of which is unique according to FAO.

Agriculture and Colonialism

- Over 2500 years ago, Indian farmers had discovered and begun farming many spices and sugarcane. It was in India, between the sixth and fourth centuries BC, that the Persians, followed by the Greeks, discovered the

famous "reeds that produce honey without bees" being grown. These were locally called, pronounced as saccharum. On their return journey, the Macedonian soldiers carried the "honey bearing reeds," thus spreading sugar and sugarcane agriculture. People in India had also invented, by about 500 BC, the process to produce sugar crystals. In the local language, these crystals were called *khanda*, which is the source of the word *candy*.

☆ Prior to 18th century, cultivation of sugar cane was largely confined to India. A few merchants began to trade in sugar - a luxury and an expensive spice in Europe until the 18th century. Sugar became widely popular in 18th-century Europe, then graduated to becoming a human necessity in the 19th century all over the world. This evolution of taste and demand for sugar as an essential food ingredient unleashed major economic and social changes. Sugarcane does not grow in cold, frost-prone climate; therefore, tropical and semitropical colonies were sought. Sugarcane plantations, just like cotton farms, became a major driver of large and forced human migrations in 19th century and early 20th century - of people from Africa and from India, both in millions - influencing the ethnic mix, political conflicts and cultural evolution of various Caribbean, South American, Indian Ocean and Pacific island nations. The history and past accomplishments of Indian agriculture thus influenced, in part, colonialism, first slavery and then slavery-like indentured labour practices in the new world, Caribbean and the world history in 18th and 19th centuries.

Early Common Era–High Middle Ages (200–1200CE)

☆ The Tamil People cultivated a wide range of crops such as rice, sugarcane, millets, black pepper, various grains, coconuts, beans, cotton, plantain, tamarind and sandalwood. Jackfruit, coconut, palm, areca and plantain trees were also known. Systematic ploughing, manuring, weeding, irrigation and crop protection was practiced for sustained agriculture. Water storage systems were designed during this period (1st-2nd century CE), a dam built on river Kaveri during this period, is considered the as one of the oldest water-regulation structures in the world still in use.

☆ Spice trade involving spices native to India–including cinnamon and black pepper–gained momentum as India starts shipping spices to the Mediterranean. Roman trade with India followed as detailed by the archaeological record and the Periplus of the Erythraean Sea Chinese sericulture attracted Indian sailors during the early centuries of the common era.

☆ Crystallised sugar was discovered by the time of the Guptas (320-550 CE), and the earliest reference of candied sugar come from India. The process was soon transmitted to China with traveling Buddhist monks. Chinese documents confirm at least two missions to India, initiated in 647

CE, for obtaining technology for sugar-refining. Each mission returned with results on refining sugar0. Indian spice exports find mention in the works of Ibn Khurdadhbeh (850), al-Ghafiqi (1150), Ishak bin Imaran (907) and Al Kalkashandi (fourteenth century).

Middle Ages

- During the Middle Ages, farmers in North Africa, the Near East, and European began making use of agricultural technologies including irrigation systems based on hydraulic and hydrostatic principles, machines such as norias, water-raising machines, dams, and reservoirs. This combined with the invention of a three-field system of crop rotation and the moldboard plow greatly improved agricultural efficiency. In the European medieval period, agriculture was considered part of the set of seven mechanical arts.
- Noboru Karashima's research of the agrarian society in South India during the Chola Empire (875-1279) reveals that during the Chola rule land was transferred and collective holding of land by a group of people slowly gave way to individual plots of land, each with their own irrigation system. The growth of individual disposition of farming property may have led to a decrease in areas of dry cultivation. The Cholas also had bureaucrats which oversaw the distribution of water - particularly the distribution of water by tank-and-channel networks to the drier areas.

Man Started Cooking 1.9 Million Years Ago

- Scientists have found evidence which suggest that early humans cooked their first hot meals more than 1.9 million years ago, much earlier than originally thought. Researchers at Harvard University who studied tooth sizes and the feeding behaviour of extinct hominids, monkeys, apes and modern humans concluded that cooking was commonplace among Homo erectus, our flat-faced and thick-browed ancestors which lived 1.9 million years ago.
- "We see a dramatic shift in the tooth size of Homo erectus, which means it was likely responding to a history already of eating cooked and processed food," said Chris Organ, an evolutionary biologist at Harvard. "If you're cooking your food you have many more hours of your day free since you don't have to eat as much to get your daily requirements," evolutionary biologist Organ said.
- Processed food is much easier to chew and digest and since chewing breaks up the food it means more surface area is available from which the gut can absorb nutrients, Organ said. The result means more available calories per serving and less gut time needed to digest those calories. The research also found that chimpanzees spent 10 times longer chewing and eating than humans do.

- It also found that Homo erectus, which emerged in Africa around 1.9 million years ago, spent 6.1 per cent of its time eating, while the Neanderthals spent 7 per cent of their timing for humans. Looking back over two million years to a more distant relative, the researchers found that Homo habilis spent about 7.2 per cent of its time eating and Homo rudolfensis 9.5 per cent.

Late Middle Ages – Early Modern Era (1200 -1757CE)

- The Muslim Farmers in North Africa and the Near East of the Medieval world are credited with inventions of extensive irrigation based on hydraulic and hydrostatic principles such as norias, water mills, water raising machines, dams and reservoirs.
- Grand Anicut dam on river Kaveri (1st -2nd Century CE) is one of the oldest water-regulation structures in the world still in use.
  - The Renaissance saw the innovation of the three field system of crop rotation and wide spread usage of the moldboard plow
  - The early phase of Industrial Revolution witnessed new agricultural practices like enclosure, mechanization, four-field crop rotation and selective breeding.

Modern Era

- After 1492, a global exchange of previously local crops and livestock breeds occurred. Key crops involved in this exchange included the tomato, maize, potato, manioc, cocoa bean and tobacco going from the New World to the Old, and several varieties of wheat, spices, coffee, and sugar cane going from the Old World to the New. The most important animal exportation from the Old World to the New were those of the horse and dog (dogs were already present in the pre-Columbian Americas but not in the numbers and breeds suited to farm work). Although not usually food animals, the horse (including donkeys and ponies) and dog quickly filled essential production roles on western-hemisphere farms.
- The potato became an important staple crop in northern Europe. Since being introduced by Portuguese in the 16th century, maize and manioc have replaced traditional African crops as the continent's most important staple food crops. By the early 19th century, agricultural techniques, implements, seed stocks and cultivated plants selected and given a unique name because of its decorative or useful characteristics had so improved that yield per land unit was many times that seen in the Middle Ages. Although there is a vast and interesting history of crop cultivation before the dawn of the 20th century, there is little question that the work of Charles Darwin and Gregor Mendel created the scientific foundation for plant breeding that led to its explosive impact over the past 150 years.

- With the rapid rise of mechanization in the late 19th century and the 20th century, particularly in the form of the tractor, farming tasks could be done with a speed and on a scale previously impossible. These advances have led to efficiencies enabling certain modern farms in the United States, Argentina, Israel, Germany, and a few other nations to output volumes of high-quality produce per land unit at what may be the practical limit. The Haber-Bosch method for synthesizing ammonium nitrate represented a major breakthrough and allowed crop yields to overcome previous constraints. In the past century agriculture has been characterized by enhanced productivy, the substitution of labour for synthetic fertilizers and pesticides, water pollution, and farm subsidies. In recent years there has been a backlash against the external environmental effects of conventional agriculture, resulting in the organic movement.
- The cereals rice, corn, and wheat provide 60 per cent of human food supply. Between 1700 and 1980, "the total rea of cultivated land worldwide increased 466 per cent " and yields increased dramatically, particularly because of selectively bred high-yielding varieties, fertilizers, pesticides, irrigation, and machinery. For example, irrigation increased corn yields in eastern Colorado by 400 to 500 per cent from 1940-1997. However, concerns have been raised over the sustainability of intensive agriculture, which has become associated with decreased soil quality in India and Asia, and tyhere has been increased concern over the effects of fertilizers and pesticides on the environment, particularly as population increases and food demand expands. The monocultures typically used in intensive agriculture increase the number of pests, which are controlled through pesticides. Integrated pest management (IPM), which "has been promoted for decades and has had some notable successes" has not significantly affected the use of pesticides because policies encourage the use of pesticides and IPM is knowledge-intensive.
- Although the "Green Revolution" significantly increased rice yields in Asia, yield increases have not occurred in the past 15-20 years. The genetic "yield potential" has increased for wheat, but the yield potential for rice has not increased since 1966, and the yield potential for maize has "barely increased in 35 years." It takes a decade or two for herbicide-resistant weeds to emerge, and insects become resistant to insecticides within about a decade. Crop rotation helps to prevent resistances. Agricultural exploration expeditions, since the late 19th century, have been mounted to find new species and new agricultural practices in different areas of the world. Two early examples of expeditions include Frank N.Meyer's fruit-and nut-collecting trip to China and Japan from 1916-1918 and the Dorsett-Morse Oriental Agricultural Exploration Expedition to China, Japan, and Korea from 1929-1931 to collect soybean germplasm to support the rise in soybean agriculture in the United States.

Chapter 9

Plant Protection through Indigenous Traditional Method

Systematic Plant Protection in Indian Agricultural History

References to plant protection are found in Vedas (Rigveda c.3700 BC Atharvaveda c.2000 BC), Kautilya's Artha-sastra (c.300 BC), Buddhist literature (c.200 BC), Krishi Parashar(c.100 BC), Sangam literature of Tamils (200 BC-100 AD), Agnipuran (c.400 AD), Brhat Samhita of Varahamira (c.600 AD), Kashyapiyakrisukti (c. 800-900 AD), Surapala's Vrikshayurveda (c.1000AD, Someshwara Deva's Manasollasa (c, 1100A D), Lokopakara) by Chavundaraya (c.1108 AD), Sarangadhara's Upavanavinoda (c.1300 AD),Viswavallabh of Chakrapani Mishra (c.1577 AD), and some documents of the medieval and pre-modern period. But Surapala c.1000 AD) has given plant protection in a very systemic manner right from seed treatment to the storage of grains.

Described by Surapala

- ✰ Diseases of all kinds of trees are stated to be of two types: internal and external. This was more than 700 years after Surapala had already done such classification. For the 'internal disorder of plants he borrowed the tridosha principle of Ayurveda, classified "internal causes as the imbalance of humors, vata, kapha and pitta; and external ones are caused by insects, cold weather/frost, scorching heat, water stress *etc.*
- ✰ Among these the diseases caused by vata are due to the land that becomes arid on account of excessive supply of dry and pungent matter which leads to thinness and crookedness of trunk, appearance of knots on trunk and trees, and the fruits become hard with less juice and sweetness. The

possible causes are underground mechanical barrier, leaf galling insects, root infecting fungi or nematodes, viruses, saline/alkaline soils.

- ✰ The diseases of kafa type occur in winter and spring if the trees are extensively watered with materials which are sweet, sour, salty, or cold. Affected trees take long time to bear fruits, show paleness, dwarfing of leaves, tastelessness, and ripe prematurely, oozing without wounds. The possible causes are fungal gummosis/rot, nutrient deficiencies or toxicities, excessive watering.
- ✰ The diseases of pitta type occur at the end of summer if threes are extensively watered with materials which are bitter, sour, salty, and strong. These diseases are characterized by yellowness of leaves, dropping of fruits, dryness, paleness of flowers and fruits, and decay. The possible causes are viruses, salinity in irrigation water, predisposal of blossoms blight and fruit decays due to fungal/bacterial infection.

Treatments of Ailments Suggested by Surapala

- ✰ The Ddiseases caused by vata can be cured by flesh, marrow, and ghee; sprinkling of kunapa water and liberal fumigation of the mixture of fat of hog, oil of Gangetic porpoise, ghee, hemp, hair of horses and cow's horn-all boiled and set to a decoction. Likewise diseases caused by kafa humor can be cured with bitter, strong and astringent decoction made out of Panchamula (roots of five plant species-sriphala, sarvatobhadra, patala, ganikarika and synoka) with fragrant water or the paste of white mustard should be deposited at the root and trees should be watered with a mixture of sesame and ashes. In case trees are affected by the kafa disease, soil around the roots of the tree should be removed and fresh dry soil should be replaced for curing them. To get cured from pitta type of diseases, treat trees with cool and sweet substances-when watered by decoction of milk, honey, yastimadhu (*Glycyrrhiza glabra* Linn.) and madhuka (*Madhuka indica* J. F. Gmel). Further, when watered with the decoction of fruits of triphala (the three kinds of myrobalan-*Terminalia chebula* Retz, *T. belliraca* Roxb., and *Phyllanthus emblica* Linn.), ghee, and honey the trees are freed of all diseases arising from the state of ailments..
- ✰ To remove insects both from the roots and branches of the trees, water the trees with cold water for seven days. The worms population can be over come by the paste of kunapa jal (water) and cow dung mixed with water and also by smearing the roots with a mixture of white mustard, vaca (*Zingiber zerumbet* Rosc. Ex Smith.), kusta (*Saussurea lappa* C. B. Clarke), and ativisa (*Aconitum heterophyllum* Wall ex Royle). Likewise the insects on the leaves can be destroyed by sprinkling the powder of ashes and dust. A wound caused by insects heals if sprinkled with milk after being anointed with a mixture of vidanga, sesame, cow's urine, ghee, and mustard. Other wounds of the trees are healed by the treatment of

anointing with the paste of bark of nyagrodha (*Ficus bengalensis* Linn.), and udumbara (*F. glomerata* Roxb.), cow dung, honey, and ghee. The oozing can be cured by the use of above paste and covering the part with the bark of dhava (*Anogeissus latifolia*), sriparnika (*Myrica esculenta* Buch-Ham.), syama (*Ichnocarpus frutescens* R. Br.), vetasa (*Salix capera* Linn.) and arjuna (*Terminalia arguna* (Roxb.) Wight and Arn.

- Similarly Surapala had suggested the different treatments like sprinkling of kunapa water and milk, anointing the branches of trees with vidari (*Pueraria tuberose* DC), sugar, nagajivha (red arsenic) and sesame mixed together and sprinkled with milk water to the trees suffering from frost, scorching heat; and if branches burnt, respectively. Further, if the trees dried due to bad soil, the original soil from the root should be removed and it should be replaced by healthy soil and milk- water should be sprinkle on it. If the drying is due to lack of water the trees should be watered with milk-water and properly fomented by smoke of the crab shells.
- Diseases caused by wrong treatment can be conquered by sprinkling the mixture of water and milk and also by applying a paste of vidinga mixed with thick mud. Jaundice (yellowing) can be brought under control only in weeks by sprinkling water mixed with the powder of barley and wheat added to honey and milk.

Seed Procurement

- Surapala in Vrikshayurveda explain the method of procurement of seed. According to him the seeds should be extracted from dried fruits which become ripe in natural course and season. It is sprinkled with milk and dried for five days, then smoked with mustard (*Brassica juncea* L.) seeds with vidanga (*Embelia ribes* Burm. F). Milk is sprinkled to protect the seeds from viral infection and mustard is having anti–insect properties due to presence of sinalbin, nematicidal properties due to glucosinolates and antifungal activity due to allyl isothiocyanate. Vidanga is antibacterial and insecticidal due to presence of embelin (benxaquinone) which is effective against stored grain pests.
- Therefore sSmoking of seeds with mustard and vidanga protects the seeds from various fungal diseases and pest attack and at the same time, the treatment induces disease resistance after germination of the seeds.

Infectious Diseases

- Surapala suggested that before planting cuttings in the pits, the latter should be 'burned' using dry plant material, cow dung *etc*. This is an indication of a suspicion that Surapala must have known about existence of infectious entities. Chakrapani Mishra (1577 AD) suggested that diseased plants found in the midst of healthy plants should be removed and burned, this again pointing towards existence of infectious entities.

It is unfortunate that all current textbook on plant pathology credit Tillet, who in 1755 AD dusted wheat seed with 'bunt' spores to produce the disease called wheat bunt. We should, however know that Koch's postulates have to be followed to prove infectious nature of a disease, Here again, Indians have not been given due credit by the authors of the West.

- Vishavavallabha is an another treatise written by Chakrapani Mishra under the patronage of Maharana Pratap of Mewar on the science of plant life which resembles Surpala's Vrikshayurveda and deals more or less with the same subject but with some additions. For example, several new herbs have been mentioned for the control of disorders, such plant species are, ambu (*Pavonia odorata* Willd.), aragavadha (*Cassia fistula* L.), arishta (*Sapindus emarginatus* Vahl.), ingundi (*Nalanites aegyptiaca* (L.) Delile), karanja (*Pongamia pinnata* (L.) Pierre), katphala (*Myrica esculenta* Buch.-Ham, ex D. Don), katvanga (*Ailanthus excelsa* Roxb.), kuberakshi (*Caesalpinia crista* L.), nimba (*Azrdirachta indica* A, Juss.) bark, rohita (*Tecomella undulate* (Smith) Seem.), shatapushpa (*Anethum sowa* Kurz.), tagara (*Valeriana jatamansi* Jones, vasa (*Adhatoda vasica* Nees), *etc.*
- Apart from Vrikshayurveda and Vishvavallabha paramount documents concerning plant protection were Someshwara Deva's Manosollasa (1131 AD), Sarangdhara's Upvanvinoda (1300 AD), Bhavprakash-nighantu (1600AD), Tuzak-i-Jahangiri (1605 AD), Dara Shikoh's Nuskha Dar-Fanni-Falahat (1650 AD), Jati Jai Chand Diary (1658-1714 AD), an anonymous Rajasthani manuscript from Mewar region of Rajasthan(1877 AD), and Watt's Dictionary of Economic Products of India (1889- 1893).
- Jahangir, the Mughal Emperor of India (1605-1627) described in his memoir "a disorder of marigold" which could be ascribed today to species of Alternaria, Botrytis or Sclerotium. Similarly in Jati Jaichand diary the early blight (Curvularia penniseti) of pearl millet and possibly Botrytis gray mold of chickpea have been described.
- In a document of early 19th century from the Mewar region of Rajasthan, powdery mildew has been described infesting various plants alongwith canker or anthracnose of orange. In this document a number of plant protection practices have been given. Some interesting practices are:
    - Use of oil (probably sesame) for soil and foliar application to trees to protect from frost and termites.
    - Sprinkling of curd (9 L) with asafetida (112 g) on trees to prevent powdery mildew.
    - Use of asafetida and vidanga mixed with curd every 10 days to protect against orange canker.
- Use of cow dung for smearing the cutting of fig before planting is mentioned in Dara Shikoh's Nuskha Dar Fanni_Falahat (1650 AD). Garlic has been mentioned specially for insect control.

- In addition to these he has mentioned the use of salt solution for soaking fig cuttings before planting. This is followed by cow dung application.

Some Indigenous Plant Protection Practices Still Followed

- In traditional agricultural practices, farmers evolved an effective system of crop protection through generations of experience and intimate knowledge of their environment which are still followed by the farmers in different parts of the country. Some of such practices are: a
- application of organic manure, summer ploughing, crop rotation, use of neem leaves and neem cake, seed treatment with ash, *etc.*
- Besides these several indigenous practices which are still followed by the farmers of Arunchal Pradesh and Rajasthan are mentioned below:
  - Buttermilk for pest control. Farmers treat garlic seeds with buttermilk @ 10-12 l/ha before sowing for protection against different diseases and insect pests.
  - Asafetida to control of ergot of sorghum. About 50-60 g of asafetida in water is used for making solution to treat sorghum seeds for protection against ergot.
  - Control of ginger rot and yellowing. The tribal farmers sow ginger after treating with a solution of cow dung. About 10-15 kg cow dung and asafetida (40-80g) are mixed in 8-10 L of water. This is enough to treat ginger rhizomes required for one hectare. This practice is helpful for protecting ginger plants from rotting during vegetative growth. The farmers grow papaya as an inter-crop for providing shade to ginger and to prevent yellowing. The leaves of ginger do not turn yellow and high yields are obtained.
  - Control of tomato wilt. Farmers use turmeric powder @ 15 to 20 kg/l of water to control wilt (*Fusarium* sp.) of tomato. This solution is used to treat the roots of seedlings grown in nurseries of tomato before transplanting.
  - Control of early stem borer in sugarcane. To control the stem borer (*Chilo infuscatellus*) in sugarcane, farmers use neem oil @ 1.5-2 L/ha. This practice is repeated thrice in the whole period of sugarcane vegetative growth.
  - Control of termites. Farmers keep asafetida in a pack of cotton cloth at two or three points in the irrigation channel of 10-15 m irrigation for controlling termites (*Odonotermes obescus*) in affected crops.
  - Control of storage pests. Farmers used different simple practices for the control of storage pests *viz.* fuel wood ash for pests of pulses (250 g ash/250 kg of pulses); dry neem leaves @ 2.5 kg/100 kg of wheat for controlling storage pests, dry chillis @ 5-6 per kg of seed of mung bean and black gram.

Ancient Crop Protection Practices and their Relevance Today

- ☆ It is believed that the first organised agriculture developed in West Asia. Some early village sites such as Jarmo in the Kurdish hills (Iraq) date back to about 5000 BC. Some scholars believed that in contrast to the field-scale agriculture of the West Asia, garden production of root crops prevailed in parts of Asia. Recent research however is showing that even cereal agriculture my have preceded in China over West Asia.
- ☆ After the beginning of agriculture, humans had to worry about the protection of plants. It began when humans attempted to understand ailments affecting crops, which now are known as 'abiotic' and 'biotic' disorders. It must also be taken for granted that the causal agents were present too. Bacteria, fungi, viruses, insects and nematodes must have affected plants for millennia. But, the crucial step towards plant protection took place when the pests invaded the first crop of cultivated plants. Therefore, we can assume that the biotic and abiotic disorders were already present when the humans made their appearance. What is important to note is that plant diseases were well recognised by ancient Indian scientists and that they developed organic agents as pesticides.
- ☆ It is now being recognised by modern agricultural science also that organic manure and pesticides are better than the chemical ones. These time tested traditional methods have become more relevant today when the effects of the so called Green Revolution are wearing off, based as it was heavily on chemical inputs. And now we are getting into the genetically modified crops, without much concern for their long term consequences.
- ☆ As far as the Indian subcontinent is concerned, there are some valuable documents available which contain some information on man's efforts to protect his crops.

Ancient crop protection are summarized and under:

1. Identification of Disorders

As mentioned earlier, man began to protect his crops as soon as he started settling on land and practice organized agriculture. Initially, his concerns were pets like birds and animals, but soon he realized that crops suffer from some other disorders also, the causes of which were not clear. As a result he turned to gods and goddesses for the protection of his crops. The disorders can be divided into two types: biotic and abiotic depending upon the causal agents living or non-living respectively. There are also some internal disorders based on an Ayurvedic concept. All of these are described below in brief.

Biotic Disorders

Biotic disorder and described as under:

- *Birds*- The earliest references about birds as pests were found in *Rigveda*. Protection of a ready to harvest crop by shouting to scare the birds has been also mentioned. The other texts mentioned earlier also refer to damage to mature crops by birds. Some specific birds mentioned are parakeets, sparrows, crows, hawks and several others.
- *Rats*- *Atharvaveda*, Kautilya's *Arthasastra* and several other texts mention rat, as a pest, attacking crops in the field as well as stored grains.
- *Other animals*- Several other animals such as wild boar, deer, goats and buffaloes *etc.* were also supposed to damage the crops.
- *Locusts and termites*- Both locusts and termites are mentioned as pests in several texts.
- *Other insects*- References to specific insects like *Leptocorisa varicornis* (Gandhi bug), *pandarmundi* (white earhead), *Tryporyza incertulas etc.* are found in the literature of 19th and 20th centuries.
- *Phanerogamic parasites*- A phanerogamic semiparasite *Loranthus longiflorus* Disr. was mentioned by Susruta (c. 400 BC). Similarly, dodder (*Cuscuta reflexa* Roxb.), a parasite has been mentioned in *Bhavaprakashanighantu* (Pandey and Chunekar, 1999).
- *Algae and fungi*- Jahangir (1605–1627), in his memoirs, described a disorder of marigold, which could be ascribed today to species of *Alternaria, Botrytis*, or *Sclerotinia*. The diseases known as "mildew of paddy" and "blight of sugarcane" were mentioned in some Buddhist documents. In the second half of the 19th century parasitism of fungi was proven in Europe. *A Dictionary of Economic Products of India* was a monumental effort of G. Watt, published during 1889–1893. This book contains a detailed description of disorders of crops covering the period from 1820s.

Abiotic Disorders

Abiotic disorder and described as under:

- Surpala, for the first time in the history of world agriculture classified plant disorders into two types: internal and external.
- The abiotic disorders were considered as the external type and are mentioned below.
    - *Scorching heat*- Due to the impact of scorching heat, the leaves start looking pale, show yellowing, and finally dry.
    - *Frost*- Crop damage due to the impact of frost is mentioned but the symptoms are not specifically described. However, we can imagine that it must ultimately be the drying of the foliage. Varahamihira had also mentioned cold weather as a factor.

- *Stormy winds*- Severe stormy winds result in the breaking of branches, uprooting and twisting of trees.
- *Fire, lightning and soil aridity*- Jahangir, in his memoirs, has given a detailed account of damage due to lightning strike near Jalandhar in Punjab in 1621 AD. These factors result in the drying up of the trees.

> *A thorough study of Surapala reveals holistic crop management through Vrikshayurveda. Surapala in his text has stressed use of suitable cultivars, use of good seeds, pre-sowing treatment of seeds, use of suitable soils, growing intercrops, having optimum plant population, balanced nutrition, optimum use of water, timely weeding, protection from disorders using herbal products or dead animal wastes, harvest at the right stage, and seed drying and storage. We should be proud of the fact that this knowledge base existed in India about 1000 years ago.*

- *Accidental mechanical wounds*- These types of wounds (*e.g.*, with axe) may lead to drying of trees.
- *Excessive water and continuous shade*- Excessive water results in foul smell, reduction in leaf size, and stunting of plant, while continuous shade results in the destruction of trees.

Internal Disorders Based on an Ayurvedic Concept

- Ayurveda, the Indian indigenous system of medicine has been an integral part of the Indian culture. The term Ayurveda has been taken from Sanskrit, the meaning of which is knowledge of life. As the name suggests, Ayurveda is not only a science of treatment of the ill but also covers the whole gamut of happy human life.
- In functional terms, Ayurveda recognizes 3 different biological systems- *vata, pitta* and *kapha*. *Vata* controls all the movements in the body while *pitta* takes care of chemical reactions and biosynthesis of various compounds within the body. *Kapha* deals with balanced growth, development and functioning of the body. If the functions of these three humors are well balanced, the individual is supposed to be in good condition, while an imbalance within or between them, results in the imbalance leading to various kinds of ailments. This concept of Ayurveda is known as *tridoshasiddhanta*. So we can say that the primary purpose of Ayurveda is to help people to maintain *vata, pitta* and *kapha* in balanced condition to prevent from various kinds of diseases. These concepts were extrapolated to plants as well.
- Varahamihira (600 AD) wrote a chapter on treatment of trees, in which he mentioned that trees are vulnerable to diseases when exposed to cold weather, strong winds and hot sun. As a result their leaves become pale white, sprouts scanty and unhealthy, branches dry and their sap oozes

out. Surpala (1000 AD) compiled a text called *Vrikshayurveda* (vriksha= trees; *Ayurveda*= science) and described the concept as applied to trees. It appears that the basic knowledge of *Vrikshayurveda* remained mostly confined to scholars.

- The concept of *tridoshasiddhanta* in diagnosing and treating tree disorders continued to be evident in the Indian literature at least until the 16th century AD when Chakrapani Mishra wrote *Viswavallabha*.

2. Protection/Treatment Practices

Treatment Based on Mantras

Atharvaveda give references about chanting of mantras to protect crops and grains from insects such as grasshoppers and animals such as rats. Parashara (400 BC) specifically gives the following mantra for controlling grain destroyer:

> *"Salutations to the feet of the revered preceptor. Let success prevail! The evervictorious feet of Ram a (i.e., Rama himself). the Lord of Lords, the Emperor of Emperors, the revered One, commands from his heavenly abode situtated on the peak of the Himalayas, the slope of which are white like the conch, the jasmine flower, the Moon-Hanuman, the son of Wind, moving fast like wind, destroyer of invaders, standing on the seashore amidst hundreds and thousands of monkeys with his tail raised and claws harsh and strong, 'Let there be well being.' Winds are blowing with great force in a section of a farm belonging to so and so hailing from such and such family/group. If the destroyers of crops such as gandhi, shankhi, pandarmundi, dhuli, shringari, kumari, madaka, etc. and goats, wild boars, pigs, deer, buffaloes, parrots, sparrows, winged insects, etc. do not leave that farm by you order, you shall strike them hardwith your tail strong like thunderbolt."*

The mantra had to be written with the red lac-dye on a leaf and tied in the field. Surpala also described the same mantra with some variation in *Vrikshayurveda*.

Practices Using Organic Materials

- Kautilya's *Arthasastra* was probably the oldest document, which described the use of organic materials to control the crop disorders. Varahamihira mentioned the use of milk, ghee, and cow dung for dressing the seeds and smoking them by burning animal flesh or turmeric before sowing.
- Surpala mentioned various plant protection practices, some of which are as follows:
  - Sprinkling *kunapa* (liquid manure prepared from parts of carcasses) on trees suffering from imbalance of *vata*.
  - Fumigation (smoking) by burning animal fat, ghee, hemp, horse hair and cow's horn, also for *vata*.

- Sprinkling a decoction made out of *panchamula* (roots of *Clerodendrum phlomidis, Aegle marmelos, Stereospermum suaveolens, Gmelina arborea* and *Oroxylum indicum*) for *Kapha* type of disorders.
- Drenching tree base by decoction of milk, honey, licorice and *Madhuca indica* for *pitta* type of disorders.
- Drenching tree base with decoction of *triphala* (dried fruits of *Terminalia chebula*), ghee and honey; also for *pitta* type of disorders.
- Dressing tree wounds with a paste made from the barks of banyan and cluster fig trees, cow dung, honey, mustard, and ghee.
- Applying paste of *vidanga* (*Embelia ribes*) and thick mud.

- All the plant species mentioned above in these practices have biocidal properties. Honey, mustard and licorice also possess antimicrobial properties and cow dung mixed with urine also shows some medicinal properties. Even today, cow urine has been successfully used as a pesticide for growing Safed Musli (*Chlorophytum borivillanum*).
- Use of cow dung for smearing the cuttings of fig before planting is mentioned in Dara Shikoh's documents (Razia Akbar, 2000). Some interesting practices are mentioned in a 19th century document from Rajasthan (Javalia, 1999), which are described in brief as follows:
  - Use of foliar and soil applications of oil to trees to protect from frost and termites.
  - Sprinkling of curd (9 L) mixed with asafoetida (112 g) on trees to prevent powdery mildew.
  - Use of *Embelia ribes* mixed with curd every 10 days to protect canker of orange.
- *The Dictionary of the Economic Products of India* by Watt (1889-1893), also gives information about the practices followed in the 19th century in India. Some of which are as follows:
  - Application of cattle manure to pigeon pea to reduce frost damage
  - Application of leaves of *Calotropis gigantea* for two years to reclaim soils with salts efflorescing
  - Sanitation (removal of all dead organic matter from the betel leaf sheds to prevent spread of diseases)
  - Reduction of disease (collar rot) by soil application of onion juice mixed with cow dung

Practices Using Inorganic Materials

- Someshwara Deva (c.1126 AD) was the first, who suggested treatment of seed with ash, besides other material, to ensure good germination. Dara Shikoh mentioned the use of common salt solution for soaking fig cuttings prior to planting.

- Dipping seed in salt solution was a practice in the 19th century. It was Ozanne, who first described the use of copper sulphate to control sorghum smut by dipping in solution of 85 g copper sulphate in 1150 ml of water.
- The use of Bordeaux mixture (copper sulphate and lime) was first documented in India in 1906. Sulphur was also used in India in 1906-1907. The various pest controls that we are practicing today were conceptualised and practiced centuries ago.

3. Materials and Practices that Need our Early Attention

- Surpala's *Vrikshyayurveda* mentioned some materials (along with their properties) and practices that were supposed to be used in agriculture for the protection of crops. Some of these materials and practices need our attention as our agricultural scientists ignore them.
- A few of these materials are described below in brief.
    - *Application of milk and milk products:* Milk and ghee have been used for centuries. Glutamate, leucine and proline form about 40 per cent of the total amino acids in milk. Recently, a report (Arun Kumar *et al.,* 2002) claimed that milk sprays induced systematically acquired resistance in chilli against leaf curl (a viral disease). Milk also has been used for controlling powdery mildews. The amino acid proline has been found to systemically induce resistance in plants (Niranjan Raj and Shetty, 2002). High amounts of endogenous proline increase contents of cytokinin and auxins. So we can say that milk treatment requires our early attention and gives us an opportunity to rediscover its beneficial effects.
    - *Application of cow dung:* The use of cow dung has been indicated since the time of Kautilya (c. 300 BC). It was used for dressing seeds, plastering cut ends of vegetatively propagating sugarcane, dressing wounds, sprinkling diluted suspension on plants *etc.* since ancient times. Still Indian farmers use cow dung in different ways but agricultural scientists have ignored its importance. Agricultural scientists think that it can be used as manure only.
    - Cow dung is a mixture of dung and urine, generally in the ratio of 3:1. It contains crude fibre, crude protein and materials that can be obtained in nitrogen –free extracts and ether extracts. The cow dung also contains micronutrients. The urine portion of cow dung consists of nitrogen, potash, sulphur and traces of phosphorus.
    - When seed is treated with cow dung in various ways, it gets coated with cow dung residue that contains cellulose, hemi cellulose, micronutrients, metabolic nitrogen, epithelial cells from the animals, bile salt and pigment, potash, sulphur, traces of phosphorus and a large number of bacteria. This thin dry layer of residue on seed absorbs moisture from the surrounding soil to the advantage of the

seed. The presence of bacteria in cow dung plays a significant role in the development of the seed. As these cow dung bacteria have the capacity to utilize cellulose, hemi cellulose and pectin, so these can quickly colonize the area around sown seed and compete with the pathogenic fungi and bacteria and prevent them from attacking the seed. As Indian farmers are using cow dung for a long time, they are convinced of its utility. Now it is the duty of agricultural scientists to take initiative, as there is a lot to learn about the role of cow dung in maintaining the seed health. Dried cow dung powders could also be applied to soil to promote bio-control.

- Nene informs that it is good news to know that according to a newspaper report (15 March 2003) that scientists at the Indian Institute of Technology at Delhi and Kanpur have taken up a project on researching cow urine.
- ***Application of liquid manure (kunapa):*** *Kunapa's* preparation involves boiling flesh, fat and marrow of animals such as deer, pig, fish, sheep or goats in winter, placing it in an earthen pot and adding milk, powders of sesame oilcake, black gram boiled in honey, decoction of pulses, ghee and hot water. Now the pot containing these materials is put in a warm place for two weeks. The resulted fermented liquid manure is known as *kunapa*. There is always a danger of passing on dormant pathogens to fields with plant-based composts, but there should be no such danger with application of *kunapa* water.
- Firminger (1864), who was known as the "Chaplain of the Bengal Establishment" mentions beneficial use of liquid manure for vegetable cultivation, but he has given no information about who first thought of liquid manure.
- All the materials used in the preparation of *kunapa* need detailed research so that we might be able to provide acceptable scientific evidence to support recommendations made by Surpala.
- ***Application of some other materials :*** Some other materials mentioned by Surpala were animal fat, ash, brick powder, buffalo horn, cow horn, crab shells, fish meal, honey, horse hair, lotus mud, marrow *etc.* All these materials were recommended by Surpala to control tree disorders. Some plant species like *Acorus calamus* L., *Oroxylum indicum, Solanum indicum* L., *Piper nigrum* L., *Embelia ribes* Burm. F. *etc.*

Age-Old Plant Protection Practices Techniques vs Scientific Basis

- It has been now realized that the techniques adopted for commercial agriculture are unsustainable on long term basis. Therefore, agricultural scientists are diverting their attention to the traditional or indigenous technology and exploring possibilities of using them wherever possible.

- Our old traditional technologies were scientific and almost eco-friendly as all the plant protection practices were based on organic materials both of plant and animal origin which includes honey, ghee, milk and milk products, cow dung and urine, and extracts from number of plant species like *Brassica* spp., *Madhuka indica, Ficus* spp., *Piper nigrum, Azadirachta indica, Vitex nigundo, Embelia ribes etc.* The biochemical analysis of these materials clearly indicated now that all these materials have antimicrobial activities.
- Milk and ghee have been used for centuries. Even buttermilk was found useful. About 40 per cent of total amino acids in milk are glutamate, leucine, and proline. A recent report claimed that milk sprays induced systemically acquired resistance in chilli against leaf curl, a viral disease. Milk (10 per cent *acqueous suspension*) has also been effectively used for controlling powdery mildews.
- The use of cow dung by our farmers for different purposes like seed dressing, plastering cut ends of vegetatively propagating units such as sugarcane sets, dressing wounds, sprinkling dilute suspension of plants and applying to soil has been indicated since the time of Kautilya (c.300 BC). The cow dung from the cattle shed is a mixture of dung and urine in a ratio of 3:1. Cow dung consist crude fibre, crude protein and materials that can be obtained in nitrogen-free extracts. There are more than 60 species of bacteria and 100 species of protozoa encountered in the rumen of cow. Thus when seed is treated in various ways with cow dung, it gets coated with cow dung residue. The residue contains several organic elements, enzymes, macro and micro-nutrients, epithelial cells, bile salt and pigment and large number of bacteria. The dung residue has emulsifying properties and readily absorbs moisture from the surrounding soil to the advantage of seeds. The presence of bacteria may antagonize potential pathogens ready to attack seed.
- Neem cake application to the field reduces population of soil-borne fungi and nematodes and also reclaims alkaline soil due to presence of calcium and magnesium. The ancient practice of spreading of neem leaves over groundnut in storage has a scientific basis. It has now proved that neem leaves inhibit the growth of *Aspergillus flavus* and thereby prevent aflatoxin production.

Chapter 10

Crop Voyage in India and World

Introduction: An Overview

- ☆ Agriculture is a primary activity. It includes growing crops, fruits, vegetables, flowers and rearing of livestock.
- ☆ In the world, 50 per cent of persons are engaged in agricultural activity. Two-thirds of India's population is still dependent on agriculture.
- ☆ Favourable topography of soil and climate are vital for agricultural activity. The land on which the crops are grown is known as arable land in the map you can see that agricultural activity is concentrated in those regions of the world where suitable factors for the growing of crops exist.

Time Line: India and World

- ☆ Agriculture was developed at least 10,000 years ago, and it has undergone significant developments since the time of the earliest cultivation. Independent development of agriculture occurred in northern and southern China, Africa's Sahel, New Guinea and several regions of the Americas. Agricultural practices such as irrigation, crop rotation, fertilizers, and pesticides were developed long ago but have made great strides in the past century. The Haber-Bosch method for synthesizing ammonium nitrate represented a major breakthrough and allowed crop yields to overcome previous constraints.
- ☆ In the past century agriculture has been characterized by enhanced productivity, the substitution of labor for synthetic fertilizers and pesticides, selective breeding, mechanization, water pollution, and farm

subsidies. In recent years there has been a backlash against the external environmental effects of conventional agriculture, resulting in the organic movement.

1. 9500 BCE (Earliest Evidence for Domesticated Wheat)

> *Agricultural Development refers to efforts made to increase farm production in order to meet the growing demand of increasing population. Agriculture has developed at different places in different parts of the world. Developing countries with large populations usually practise intensive agriculture where crops are grown on small holdings mostly for subsistence. Larger holdings are more suitable for commercial agriculture as in USA, Canada and Australia.*

- ☆ Identifying the exact origin of agriculture remains problematic because the transition from hunter-gatherer societies began thousands of years before the invention of writing. Nonetheless, archaeobotanists/ paleoethnobotanists have traced the selection and cultivation of specific food plant characteristics, such as a semi-tough rachis and larger seeds, to just after the Younger Dryas (about 9,500 BC) in the early Holocene in the Levant region of the Fertile Crescent. There is earlier evidence for use of wild cereals: anthropological and archaeological evidence from sites across Southwest Asia and North Africa indicate use of wild grain (*e.g.*, from the ca. 20,000 BC site of Ohalo II in Israel, many Natufian sites in the Levant and from sites along the Nile in the 10th millennium BC). There is even evidence of planned cultivation and trait selection: grains of rye with domestic traits have been recovered from Epi-Palaeolithic (10,000+ BC) contexts at Abu Hureyra in Syria, but this appears to be a localized phenomenon resulting from cultivation of stands of wild rye, rather than a definitive step towards domestication.
- ☆ It isn't until after 9,500 BC that the eight so-called founder crops of agriculture appear: first emmer and einkorn wheat, then hulled barley, peas, lentils, bitter vetch, chick peas and flax. These eight crops occur more or less simultaneously on PPNB sites in the Levant, although the consensus is that wheat was the first to be sown and harvested on a significant scale.

2. 8000 BCE (Evidence for Cattle Herding)

- ☆ The archaeological record for the domestication of wild forms of cattle (Bos primigenius) indicates that the process occurred independently at least twice and perhaps three times. People kept cattle around for easy access to food, including milk, blood, and meat, and for use as load-bearers and plows. The taurine (humpless, *B. taurus*) was probably domesticated somewhere in the Fertile Crescent about 8,000 years ago. Taurine cattle were apparently traded across the planet, and appear in archaeological sites of northeastern Asia (China, Mongolia, and Korea) about 5000 years ago. Evidence for domesticated zebu (humped cattle, *B. indicus*) has been discovered at the site of Mehrgahr, in the Indus Valley of Pakistan, about 7,000 years ago.

- ☆ Scholars are divided about the likelihood of a third domestication event, in Africa. The earliest domesticated cattle in Africa have been found at Capeletti, Algeria, about 6500 BP, but Bos remains are found at African sites in what is now Egypt, such as Nabta Playa and Bir Kiseiba as long ago as 9,000 years, and they may be domesticated. If these remains were indeed domesticated, then these represent the first event of domesticating cattle.

3. 7000 BCE (Cultivation of Barley; Animals are domesticated)

- ☆ Mehrgarh, one of the most important Neolithic (7000 BC to 3200 BC) sites in archaeology, lies on the "Kachi plain of Baluchistan, Pakistan, and is one of the earliest sites with evidence of farming (wheat and barley) and herding (cattle, sheep and goats) in South Asia.
- ☆ "By 7000 BC, sowing and harvesting reached Mesopotamia and there, in the super fertile soil just north of the Persian Gulf, Sumerian ingenuity systematized it and scaled it up.

4. 6500 BCE (Cattle Domestication in Turkey)

- ☆ Remains of domesticated cattle dating to 6,500 B.C. have been found in Turkey and other sites in the Near East approach this age also. Some authorities date the domestication of cattle as early as 10,000 years ago, and others almost half that amount of time. Regardless of the time frame it is generally accepted that the domestication of cattle followed sheep, goats, pigs and dogs.

5. 6000 BCE (Indus Valley grows from Wheat to Cotton and Sugar)

- ☆ Evidences of the presence of wheat and some legumes in the 6th millennium BC have been found in the Indus Valley. Oranges were cultivated in the same millennium. The crops grown in the valley around 4000 BC were typically wheat, peas, sesame seed, barley, dates and mangoes. By 3500 BC cotton growing and cotton textiles were quite advanced in the valley. By 3000 BC farming of rice had started. Other monsoon crop of importance of the time was cane sugar. By 2500 BC, rice was an important component of the staple diet in Mohenjodaro near the Arabian Sea. By this time the Indians had large cities with well-stocked granaries.

☆ By 6000 BC farming was entrenched on the banks of the Nile River. About this time, agriculture was developed independently in the Far East, probably in China, with rice rather than wheat as the primary crop. (Rice comes up in East Asia) Archaeological investigation has identified evidence of irrigation in Mesopotamia and Egypt as far back as the 6th millennium BCE, where barley was grown in areas where the natural rainfall was insufficient to support such a crop. In the Zana Valley of the Andes Mountains in Peru, archaeologists found remains of three irrigation canals radiocarbon dated from the 4th millennium BCE, the 3rd millennium BCE and the 9th century CE. These canals are the earliest record of irrigation in the New World. Traces of a canal possibly dating from the 5th millennium BCE were found under the 4th millennium canal. Sophisticated irrigation and storage systems were developed by the Indus Valley Civilization in Pakistan and North India, including the reservoirs at Girnar in 3000 BCE and an early canal irrigation system from circa 2600 BCE. Large scale agriculture was practiced and an extensive network of canals was used for the purpose of irrigation. (Irrigation aids farming).

6. 5500 BCE (Sumerians Start Organized Agriculture)

☆ By the Bronze Age, wild food contributed a nutritionally insignificant component to the usual diet. If the operative definition of agriculture includes large scale intensive cultivation of land, mono- cropping, organized irrigation, and use of a specialized labour force, the title "inventors of agriculture" would fall to the Sumerians (The Sumerians were one of the earliest urban societies to emerge in the world, in Southern Mesopotamia), starting ca. 5,500 BC. Intensive farming allows a much greater density of population than can be supported by hunting and gathering, and allows for the accumulation of excess product for off-season use, or to sell/barter. The ability of farmers to feed large numbers of people whose activities have nothing to do with agriculture was the crucial factor in the rise of standing armies.

☆ Sumerian agriculture supported a substantial territorial expansion which along with internecine conflict between cities made them the first empire builders. Not long after, the Egyptians, powered by farming in the fertile Nile valley, achieved a population density from which enough warriors could be drawn for a territorial expansion more than tripling the Sumerian empire in area.

7. 5400 BCE (Archaelogical Proof for Domestication of Chicken)

☆ Chickens (*Gallus domestics*) were first domesticated from a wild form called red jungle fowl, a bird that still runs wild in most of Southeast Asia. It was probably domesticated by about 8,000 years ago in what is now Thailand; however, recent research suggests there may have been multiple origins in distinct areas of South and Southeast Asia.

- Genetic studies suggest that the original domesticated chicken was probably in Thailand, although multiple origin locations have been suggested as well. The first archaeological evidence to date is from China about 5400 BC, in geographically widespread sites such as Cishan (Heibei province, ca 5300 BC), Beixin (Shandong province, ca 5000 BC), and Xian (Shaanxi province, ca 4300 BC). Domesticated chickens appear at Mohenjodaro in the Indus Valley by about 2000 BC and, from there the chicken spread into Europe and Africa.

8. 5400 BCE (Linearbandkeramik Culture in Europe)

- The Linearbandkeramik Culture (LBK) is the name given by German archaeologist F. Klopfleisch in 1884 to the first true farming communities in central Europe, dated between 5400 and 4900 BC. Thus, LBK is considered the first Neolithic cultures in the European continent.
- The word Linearbandkeramik refers to the distinctive banded decoration found on pottery vessels on sites spread throughout central Europe, from south-western Ukraine and Moldova in the east to the Paris Basin in the west. The LBK people are considered the importers of agricultural products and methods, moving the first domesticated animals and plants from the Near East and Central Asia into Europe.

9. 5000 BCE (Africa Grows Rice, Sorghum)

- In China, rice and millet were domesticated by 8000 BC, followed by the beans mung, soy and azuki. In the Sahel region of Africa local rice and sorghum were domestic by 5000 BC. Local crops were domesticated independently in West Africa and possibly in New Guinea and Ethiopia.

10. 4000 BCE (Ploughs make an Appearance in Mesopotamia)

- Circa 4,000 BC, the plough (variously, plow) is believed to have been invented by the Sumerians of Mesopotamia. In its initial form, the plough would probably have been nothing more than a forked tree limb, the one prong having been sharpened in order for it to cut into the ground. The plough made it possible to harness the power of oxen to dig the furrows in which the grain seeds would be sown. And, despite the fact that most history books give the 18th century English farmer, Jethro Tull, the credit for having invented the seed drill', one has been found to be illustrated on a carved stone seal from Sumer. The seed drill was a variation of the plough, which dug the furrow, but which also contained a funnel and tube assembly to drop the seeds into the furrow at the same time.
- Though some hypothesize that Domestication of the horse occurred as early as 4000 BC in the Ukraine, the horse was definitely in use by the Sumerians around 2000 BC.

11. 3000 BCE (Maize is Domesticated in Americas)

☆ Maize was first domesticated, probably from teosinte, in the Americas around 3000-2700 BC, though there is some archaeological evidence of a much older development. The potato, the tomato, the pepper, squash, several varieties of bean, and several other plants were also developed in the New World, as was quite extensive terracing of steep hillsides in much of Andean South America. Agriculture was also independently developed on the island of New Guinea.

12. 3000 BCE (Turmeric is Harvested at Indus Valley)

☆ By 3000 B.C. turmeric, cardamom, pepper and mustard were harvested earlier in India.

13. 2737 BCE Tea is Discovered

☆ The history of tea in China is long and complex. The Chinese have enjoyed tea for millennia. Scholars hailed the brew as a cure for a variety of ailments; the nobility considered the consumption of good tea as a mark of their status, and the common people simply enjoyed its flavor. Tea was first discovered by the Chinese Emperor Shennong in 2737 BC. It is said that the emperor liked his drinking water boiled before he drank it so it would be clean, so that is what his servants did. One day, on a trip to a distant region, he and his army stopped to rest. A servant began boiling water for him to drink, and a dead leaf from the wild tea bush fell into the water. It turned a brownish color, but it was unnoticed and presented to the emperor anyway. The emperor drank it and found it very refreshing, and cha (tea) was born. While historically the origin of tea as a medicinal herb useful for staying awake is unclear, China is considered to have the earliest records of tea drinking, with recorded tea use in its history dating back to the first millennium BC. The Han Dynasty used tea as medicine. The use of tea as a beverage drunk for pleasure on social occasions dates from the Tang Dynasty or earlier.

14. 2000 BCE 1st Windmill in Babylon

☆ The first true windmill, a machine with vanes attached to an axis to produce circular motion, may have been built as early as 2000 B.C. in ancient Babylon. By the 10th century A.D., windmills with wind-catching surfaces as long as 16 feet and as high as 30 feet were grinding grain in the area now known as eastern Iran and Afghanistan.

☆ The western world discovered the windmill much later. The earliest written references to working wind machines date from the 12th century. These too were used for milling grain. It was not until a few hundred years later that windmills were modified to pump water and reclaim much of Holland from the sea.

15. 1000 BCE Sugar Processing in India

- ☆ Sugar cane originated in New Guinea where it has been known since about 6000 BC. From about 1000 BC its cultivation gradually spread along human migration routes to Southeast Asia and India and east into the Pacific. It is thought to have hybridized with wild sugar canes of India and China, to produce the 'thin' canes. It spread westwards to the Mediterranean between 600-1400 AD. Sugar cane has a very long history of cultivation in the Indian sub-continent. The earliest reference to it is in the Atharva Veda (c. 1500-800 BC) where it is called ikshu and mentioned as an offering in sacrificial rites. The Atharva Veda uses it as a symbol of sweet attractiveness.
- ☆ Sugar cane was originally grown for the sole purpose of chewing, in southeastern Asia and the Pacific. The rind was removed and the internal tissues sucked or chewed. Production of sugar by boiling the cane juice was first discovered in India, most likely during the first millennium BC. The word 'sugar' is thought to derive from the ancient Sanskrit sharkara. By the 6th century BC sharkara was frequently referred to in Sanskrit texts which even distinguished superior and inferior varieties of sugarcane. The Susrutha Samhita listed 12 varieties; the best types were supposed to be the vamshika with thin reeds and the paundraka of Bengal. It was also being called guda, a term which is still used in.

16. 500 BCE Row Cultivation in China

- ☆ The greatest achievement in the field of agriculture is row cultivation and intensive hoeing. In Europe, as with the rest of the world, they practiced scatter seed farming. Scatter seed farming is the practice of throwing the seed onto the fields at random. By throwing the seed randomly, half the seeds would not grow and make it impossible to weed the field.
- ☆ The Chinese on the other hand, planted individual seeds and rows, thus reducing seed loss. The planting of crops in rows also allowed for intensive hoeing, which in turn reduce weeds.

17. Year 200 (Multi-tube Seed Drill Invented in China)

- ☆ A seed drill is a device allowing to plant seeds in the soil. Before the introduction of seed drill, the common practice was to "broadcast" seeds by hand. Besides being wasteful, broadcasting was very imprecise and led to a poor repartition of seeds; leading to low productivity.
- ☆ The Sumerians used primitive single-tube seed drills around 1,500 BCE, but the invention never reached Europe. Multi-tube seed drills were invented by the Chinese in the 2nd century BCE.

18. Year 700 (Arab Agriculture Revolution)

- The Islamic Golden Age from the 8th century to the 13th century witnessed a fundamental transformation in agriculture known as the Arab Agricultural Revolution, or Medieval Green Revolution. The global economy established by Muslim traders across the Old World, enabled the diffusion of many crops and farming techniques among different parts of the Islamic world, as well as the adaptation of crops and techniques from beyond the Islamic world. Crops from Africa such as sorghum, crops from China such as citrus fruits, and numerous crops from India such as mangoes, rice, and especially cotton and sugar cane, were distributed throughout Islamic lands, which previously had not grown these crops.
- Some writers have referred to the diffusion of numerous crops during this period as the Globalisation of crops. These introductions, along with an increased mechanization of agriculture (see Industrial growth below), led to major changes in economy, population distribution, vegetation cover, agricultural production and income, population levels, urban growth, the distribution of the labour force, linked industries, cooking and diet and clothing in the Islamic world.

19. Year 1000 (Coffee Originates in Arabia)

- Coffee as we know it kicked off in Arabia, where roasted beans were first brewed around A.D.1000. By the 13th century Muslims were drinking coffee religiously. The bean broth drove dervishes into orbit, kept worshipers awake, and splashed over into secular life. And wherever Islam went, coffee went too.

20. Year 1492 (Columbian Exchange Changes Agriculture)

- The Columbian Exchange has been one of the most significant events in the history of world ecology, agriculture, and culture. The term is used to describe the enormous widespread exchange of plants, animals, foods, human populations (including slaves), communicable diseases, and ideas between the Eastern and Western hemispheres that occurred after 1492. Many new and different goods were exchanged between the two hemispheres of the Earth, and it began a new revolution in the Americas and in Europe.
- In 1492, Christopher Columbus' first voyage launched an era of large-scale contact between the Old and the New World that resulted in this ecological revolution: hence, it was named as "Columbian" Exchange. The Columbian Exchange greatly affected almost every society on earth, bringing destructive diseases that depopulated many cultures, and also circulating a wide variety of new crops and livestock that, in the long term, increased rather than diminished the world human population. Maize and potatoes became very important crops in Eurasia by the 1700s.

Peanuts and manioc flourished in tropical Southeast Asian and West African soils that otherwise would not produce large yields or support large populations.

21. Year 1599 1st Practical Greenhouse is Created

- The idea of growing plants in environmentally controlled areas has existed since Roman times. The Roman emperor Tiberius ate a cucumber-like vegetable daily. The Roman gardeners used artificial methods (similar to the greenhouse system) of growing to have it available for his table every day of the year. Cucumbers were planted in wheeled carts which were put in the sun daily, and then taken inside to keep them warm at night. The cucumbers were stored under frames or in cucumber houses glazed with either oiled cloth known as "specularia" or with sheets of mica, according to the description by Pliny the Elder. The first modern greenhouses were built in Italy in the thirteenth century to house the exotic plants that explorers brought back from the tropics. They were originally called giardini botanici (botanical gardens). The concept of greenhouses soon spread to the Netherlands and then England, along with the plants. Some of these early attempts required enormous amounts of work to close up at night or to winterize. There were serious problems with providing adequate and balanced heat in these early greenhouses. Jules Charles, a French botanist, is often credited with building the first practical modern greenhouse in Leiden, Holland to grow medicinal tropical plants. Originally on the estates of the rich, with the growth of the science of botany greenhouses spread to the universities.
- In the nineteenth Century the largest greenhouses were built. The conservatory at Kew Gardens in England is a prime example of the Victorian greenhouse. Although intended for both horticultural and non-horticultural exhibition these included London's Crystal Palace, the New York Crystal Palace and Munich's Gkaspalats Joseph Paxton, who had experimented with glass and iron in the creation of large greenhouses as the head gardener at Chatsworth, in Derbyshire, working for the Duke of Devonshire, designed and built the first, London's Crystal Palace. A major architectural achievement in monumental greenhouse building was the Royal Greenhouses of Laeken (1874-1895) for King Leopold II of Belgium. In Japan, the first greenhouse was built in 1880 by Samuel Cocking, a British merchant who exported herbs.

22. Year 1700 (British Agricultural Revolution)

- The 1st British Agricultural Revolution describes a period of agricultural development in Britain between the 18th century and the end of the 19th century, which saw a massive increase in agricultural productivity and net output. This in turn supported unprecedented population growth,

freeing up a significant per centage of the workforce, and thereby helped drive the Industrial Revolution. How this came about is not entirely clear.

- In recent decades, enclosure, mechanization, four-field crop rotation, and selective breeding have been highlighted as primary causes, with credit given to relatively few individuals.

23. Year 1700 (Charles Townshend Popularizes)

- Growing the same crop repeatedly on the same land eventually depletes the soil of different nutrients. Farmers avoided a decrease in soil fertility by practicing crop rotation. Different plant crops were planted in a regular sequence so that the leaching of the soil by a crop of one kind of nutrient was followed by a plant crop that returned that nutrient to the soil.
- Crop rotation was practiced in ancient Roman, African, and Asian cultures. During the Middle Ages in Europe, a three-year crop rotation was practiced by farmers rotating rye or winter wheat in year one, followed by spring oats or barley in the second year, and followed by a third year of no crops. In the 18th century, British agriculturalist Charles Townshend aided the European agricultural revolution by popularizing a four- year crop rotation with rotations of wheat, barley, turnips, and clover. In the United States, George Washington Carver brought his science of crop rotation to the farmers and saved the farming resources of the south.

24. 14/3/1794 (Cotton Gin is Invented)

- A Cotton Gin (short for cotton engine) is a machine that quickly and easily separates the cotton fibers from the seedpods and the sometimes sticky seeds, a job previously done by slave workers. These seeds were either used again to grow more cotton or if badly damaged were disposed of. It uses a combination of a wire screen and small wire hooks to pull the cotton through the screen, while brushes continuously remove the loose cotton lint to prevent jams. The term "gin" is an abbreviation for engine, and means "device". According to Joseph Needham a precursor to the cotton gin was present in India, which was known as a charkhi, which had two elongated worms that turned its rollers in opposite directions.
- The modern cotton gin was later created by the American inventor Eli Whitney in 1793 to mechanize the production of cotton fiber. The invention was granted a patent on March 14, 1794. The cotton gin was credited for increasing assets in the American economy.

25. Year 1800 (Chemical Fertilizer began to be Used)

- Chemist Justus von Liebig contributed greatly to the advancement in the understanding of plant nutrition. His influential works first denounced the vitalist theory of humus, arguing first the importance of ammonia, and later the importance of inorganic minerals. Primarily his work succeeded in setting out questions for agricultural science to address over the next 50 years. In England he attempted to implement his theories commercially through a fertilizer created by treating phosphate of lime in bone meal with sulphuric acid. Although it was much less expensive than the guano that was used at the time, it failed because it was not able to be properly absorbed by crops. At that time in England Sir John Bennet Lawes was experimenting with crops and manures at his farm at Harpenden and was able to produce a practical super phosphate in 1842 from the phosphates in rock and coprolites. Encouraged, he employed Sir Joseph Henry Gilbert, who had studied under Liebig at the University of Giessen, as director of research. To this day, the Rothamsted research station that they founded still investigates the impact of inorganic and organic fertilizers on crop yields.

- ☆ In France, Jean Baptiste Boussingault pointed out that the amount of nitrogen in various kinds of fertilizers is important. Metallurgists Percy Gilchrist and Sidney Gilchrist Thomas invented the Thomas-Gilchrist converter, which enabled the use of high phosphorus acidic Continental ores on steelmaking. The dolomite lime lining of the converter turned in time into calcium phosphate, which could be used as fertilizer known as Thomas- phosphate. In the early decades of the 20th Century the Nobel prize-winning chemists Carl Bosch of IG Farben and Fritz Haber developed the process that enabled nitrogen to be cheaply synthesized into ammonia, for subsequent oxidization into nitrates and nitrites. In 1927 Erling Johnson developed an industrial method for producing nitro phosphate, also known as the Odda process after his Odda Smelteverk of Norway. The process involved acidifying phosphate rock (from Nauru and Banaba Islands in the southern Pacific Ocean) with nitric acid to produce phosphoric acid and calcium nitrate which, once neutralized, could be used as a nitrogen fertilizer.

26. Year 1837 (John Deere Invents Steel Plough)

- ☆ In 1837 John Deere developed and manufactured the first commercially-successful cast-steel plow. The wrought-iron framed plow had a polished steel share which made it ideal for the tough soil of the Midwest, and worked better than other plows.
- ☆ By early 1838 Deere completed his first steel plow and sold it to a local farmer, Lewis Crandall, who quickly spread word of his success with Deere's plow, and so two neighbors soon placed orders with Deere. Confident that he had some stability, Deere moved his family to Grand Detour later that year. By 1841 he was manufacturing 75 plows per year and 100 plows per year the next.

27. Year 1860 (Hay Cultivation Changes)

☆ Until the middle of the 19th century, hay was cut by hand with sickles and scythes. In the 1860s early cutting devices were developed that resembled those on reapers and binders; from these came the modern array of fully mechanical mowers, crushers, windrowers, field choppers, balers, and machines for palletizing or wafering in the field. The stationary baler or hay press was invented in the 1850's and did not become popular until the 1870's. The "pick up" baler or square baler was replaced by the round baler around the 1940's. In 1936, a man named Innes, of Davenport, Iowa, invented an automatic baler for hay. It tied bales with binder twine using Appleby-type knitters from a John Deere grain binder. A Pennsylvania Dutchman named Ed Nolt built his own baler, salvaging the twine knotters from the Innes baler. Both balers did not work that well.

☆ According to The History of Twine, "Nolt's innovative patents pointed the way by 1939 to the mass production of the one-man automatic hay baler. His balers and their imitators revolutionized hay and straw harvest and created a twine demand beyond the wildest dreams of any twine manufacturer."

28. Year 1866 (Gregor Mendel Describes Mendelian Inheritance)

☆ Mendelian inheritance (or Mendelian genetics or Mendelism) is a set of primary tenets relating to the transmission of hereditary characteristics from parent organisms to their children; it underlies much of genetics. They were initially derived from the work of Gregor Mendel published in 1865 and 1866 which was "re-discovered" in 1900, and were initially very controversial. When they were integrated with the chromosome theory of inheritance by Thomas Hunt Morgan in 1915, they became the core of classical genetics.

29. Year 1879 (Milking Machine Replaces Hand Milking)

☆ In 1879, Anna Baldwin patented a milking machine that replaced hand milking - her milking machine was a vacuum device that connected to a hand pump. This is one of the earliest American patents; however, it was not a successful invention. Successful milking machines appeared around 1870. The earliest devices for mechanical milking were tubes inserted in the teats to force open the sphincter muscle, thus allowing the milk to flow. Wooden tubes were used for this purpose, as well as feather quills. Skillfully made tubes of pure silver, gutta percha, ivory, and bone were marketed in the mid-19th century.

30. Year 1892 (First Practical Gasoline-Powered Tractor)

☆ The first powered farm implements in the early 1800s were portable engines – steam engines on wheels that could be used to drive mechanical

farm machinery by way of a flexible belt. Around 1850, the first traction engines were developed from these, and were widely adopted for agricultural use. Where soil conditions permitted, like the US, steam tractors were used to direct- haul ploughs, but in the UK, ploughing engines were used for cable-hauled ploughing instead. Steam-powered agricultural engines remained in use well into the 20th century, until reliable internal combustion engines had been developed.

☆ In 1892, John Froelich built the first practical gasoline-powered tractor in Clayton County, Iowa. Only two were sold, and it was not until 1911, when the Twin City Traction Engine Company developed the design, that it became successful. In Britain, the first recorded tractor sale was the oil-burning Hornsby-Ackroyd Patent Safety Oil Traction engine, in 1897. However, the first commercially successful design was Dan Albone's three-wheel Ivel tractor of 1902. In 1908, Saundersons of Bedford introduced a four-wheel design, and went on to become the largest tractor manufacturer outside the USA.

31. Year 1900 (Birth of Industrial Agriculture)

☆ Industrial agriculture is a form of modern farming that refers to the industrialized production of livestock, poultry, fish, and crops. The methods of industrial agriculture are techno scientific, economic, and political. They include innovation in agricultural machinery and farming methods, genetic technology, techniques for achieving economies of scale in production, the creation of new markets for consumption, the application of patent protection to genetic information, and global trade. These methods are widespread in developed nations and increasingly prevalent worldwide. Most of the meat, dairy, eggs, fruits, and vegetables available in supermarkets are produced using these methods of industrial agriculture.

☆ The birth of industrial agriculture more or less coincides with that of the Industrial Revolution in general. The identification of nitrogen and phosphorus as critical factors in plant growth led to the manufacture of synthetic fertilizers, making possible more intensive types of agriculture. The discovery of vitamins and their role in animal nutrition, in the first two decades of the 20th century, led to vitamin supplements, which in the 1920s allowed certain livestock to be raised indoors, reducing their exposure to adverse natural elements. The discovery of antibiotics and vaccines facilitated raising livestock in larger numbers by reducing disease. Chemicals developed for use in World War II gave rise to synthetic pesticides. Developments in shipping networks and technology have made long-distance distribution of agricultural produce feasible. Agricultural production across the world doubled four times between 1820 and 1975 to feed a global population of one billion human beings in 1800 and 6.5

billion in 2002. During the same period, the number of people involved in farming dropped as the process became more automated.

32. Year 1930 (First Aerial Photos for Agriculture)

- ☆ Although aerial photographs were taken from balloons and kites as early as the mid-1800s, aerial survey was not widely employed until World War I (1914-1918), when cameras were mounted in airplanes. Military applications of aerial photography expanded during World War II (1939-1945), and many technological improvements in aircraft, cameras, and films followed. During the 1930s and the 1940s, the first aerial surveys of large areas of the United States were conducted to support government programs in soil conservation and forest management.

33. Year 1930 (First Plant Patent is given)

- ☆ Since 1930, plants have been patentable. The first plant patent was granted to Henry F. Bosenberg for a climbing or trailing rose.

34. Year 1939 (DDT becomes a rage)

- ☆ Since before 2500 BC, humans have utilized pesticides to protect their crops. The first known pesticide was elemental sulfur dusting used in Sumeria about 4,500 years ago. By the 15th century, toxic chemicals such as arsenic, mercury and lead were being applied to crops to kill pests. In the 17th century, nicotine sulfate was extracted from tobacco leaves for use as an insecticide. The 19th century saw the introduction of two more natural pesticides, pyrethrum which is derived from chrysanthemums and rotenone which is derived from the roots of tropical vegetables.
- ☆ In 1939, Paul Muller discovered that DDT was a very effective insecticide. It quickly became the most widely-used pesticide in the world. In the 1940s manufacturers began to produce large amounts of synthetic pesticides and their use became widespread. Some sources consider the 1940s and 1950s to have been the start of the "pesticide era." Pesticide use has increased 50 - fold since 1950 and 2.5 million tons (2.3 million metric tons) of industrial pesticides are now used each year. Seventy-five per cent of all pesticides in the world are used in developed countries, but use in developing countries is increasing. In the 1960s, it was discovered that DDT was preventing many fish-eating birds from reproducing, which was a serious threat to biodiversity. Rachel Carson wrote the best-selling book Silent Spring about biological magnification.
- ☆ DDT is now banned in at least 86 countries, but it is still used in some developing nations to prevent malaria and other tropical diseases by killing mosquitoes and other disease-carrying insects.

35. Year 1944 (Green Revolution begins in Mexico)

- ☆ The Green Revolution is the ongoing transformation of agriculture that led in some places to significant increases in agricultural production between the 1940s and 1960s. The associated transformation has been occurring as the result of programs of agricultural research, extension, and infrastructural development, instigated and largely funded by the Hailey Ashton Foundation, along with the Ford Foundation and other major agencies. The consensus among some agronomists is that the Green Revolution allowed food production to keep pace with worldwide population growth. The Green Revolution has had major social and ecological impacts, and with multi-million dollar backing from organizations including the Gates Foundation, the deployment of Green Revolution policies will continue for some time.
- ☆ The Green Revolution began in 1943 with the establishment of the Office of Special Studies, which was a venture that was collaboration between the Rockefeller Foundation and the presidential administration of Manuel Avila Camacho in Mexico. While Camacho's predecessor Cárdenas promoted peasant subsistence agriculture through policies of land reform, Avila Camacho's primary goal for Mexican agriculture was to aid in the nation's industrial development and economic growth. With the experience of agricultural development judged as a success, the Rockefeller Foundation sought to spread the Green Revolution to other nations. The Office of Special Studies in Mexico became an informal international research institution in 1959, and in 1963 it formally became CIMMYT, The International Maize and Wheat Improvement Center. In 1961 India was on the brink of mass famine.
- ☆ Norman Borlaug is known as father of green revolution. Norman Borlaug was invited to India by the adviser to the Indian Minister of Agriculture M. S. Swaminathan. Despite bureaucratic hurdles imposed by India's grain monopolies, the Ford Foundation and Indian government collaborated to import wheat seed from CIMMYT. Punjab was selected by the Indian government to be the first site to try the new crops because of its reliable water supply and a history of agricultural success. India began its own Green Revolution program of plant breeding, irrigation development, and financing of agrochemicals.

36. Year 1972 (Organic Movement Starts taking Roots)

- ☆ The organic movement broadly refers to the organizations and individuals involved worldwide in the promotion of organic farming, which they believe to be a more sustainable mode of agriculture. Its history goes back to the first half of the 20th century, when modern large-scale agricultural practices began to appear.

- The organic movement began in the early 1900s in response to the shift towards synthetic nitrogen fertilizers and pesticides in the early days of industrial agriculture. It lay dormant for many years, kept alive by a relatively small group of ecologically minded farmers. These farmers came together in various associations: Demeter International of Germany, which encouraged biodynamic farming and began the first certification program, the Soil Association of the United Kingdom, and Rodale Press in the United States, along with others. In 1972 these organizations joined to form the International Federation of Organic Agriculture Movements (IFOAM).
- In recent years, environmental awareness has driven demand and conversion to organic farming. Some governments, including the European Union, have begun to support organic farming through agricultural subsidy reform. Organic production and marketing have grown at a fast pace.

37. 1996 (Commercial Cultivation of Genetically Modified Plants)

- Transgenic plants have been developed for various purposes: resistance to pests, herbicides or harsh environmental conditions; improved shelf life; increased nutritional value - and many more. Since the first commercial cultivation of GM plants in 1996, GM plant events tolerant to the herbicides glufosinate or glyphosate and events producing the *Bt* toxin, an insecticide, have dominated the market.
- Recently, a new generation of GM plants promising benefits for consumers and industry purposes is becoming ready to enter the markets. Since GM plants are grown on open fields, there is often a perception that there could be associated environmental risks. Therefore, most countries require bio-safety studies prior to the approval of a new GM plant event, usually followed by a monitoring programme to detect environmental impacts. Especially in Europe, the coexistence of GM plants with conventional and organic crops has raised many concerns. Since there is separate legislation for GM crops and a high demand from consumers for the freedom of choice between GM and non-GM foods, measures are required to separate GM, conventional and organic plants and derived food and feed.
- European research programmes such as Co-Extra, Transcontainer and SIGMEA are investigating appropriate tools and rules. On the field level, these are biological containment methods, isolation distances and pollen barriers.

Farm System

Agriculture or farming can be looked at as a system. The important inputs are seeds, fertilisers, machinery and labour. Some of the operations involved are ploughing, sowing, irrigation, weeding and harvesting. The outputs from the system include crops, wool, dairy and poultry products.

Types of Farming

- ☆ Farming is practised in various ways across the world. Depending upon the geographical conditions, demand of produce, labour and level of technology, farming can be classified into two main types. These are subsistence farming and commercial farming.
- ☆ **Subsistence Farming:** This type of farming is practised to meet the needs of the farmer's family. Traditionally, low levels of technology and household labour are used to produce on small output. Subsistence farming can be further classified as intensive subsistence and primitive subsistence farming.
- ☆ In **intensive subsistence agriculture** the farmer cultivates a small plot of land using simple tools and more labour. Climate with large number of days with sunshine and fertile soils permit growing of more than one crop annually on the same plot. Rice is the main crop. Other crops include wheat, maize, pulses and oilseeds. Intensive subsistence agriculture is prevalent in the thickly populated areas of the monsoon regions of south, southeast and east Asia.
- ☆ **Primitive subsistence agriculture** includes shifting cultivation and nomadic herding.
- ☆ **Shifting cultivation** is practised in the thickly forested areas of Amazon basin, tropical Africa, parts of southeast Asia and Northeast India. These are the areas of heavy rainfall and quick regeneration of vegetation. A plot of land is cleared by felling the trees and burning them. The ashes are then mixed with the soil and crops like maize, yam, potatoes and cassava are grown. After the soil loses its fertility, the land is abandoned and the cultivator moves to a new plot. Shifting cultivation is also known as 'slash and burn' agriculture.
- ☆ **Nomadic herding** is practised in the semi-arid and arid regions of Sahara, Central Asia and some parts of India, like Rajasthan and Jammu and Kashmir. In this type of farming, herdsmen move from place to place with their animals for fodder and water, along defined routes. This type of movement arises in response to climatic constraints and terrain. Sheep, camel, yak and goats are most commonly reared. They provide milk, meat, wool, hides and other products to the herders and their families.

Commercial Farming

- ☆ In commercial farming crops are grown and animals are reared for sale in market. The area cultivated and the amount of capital used is large. Most of the work is done by machines. Commercial farming includes commercial grain farming, mixed farming and plantation agriculture. In commercial grain farming crops are grown for commercial purpose. Wheat and maize are common commercially grown grains. Major areas where commercial grain farming is pracised are temperate grasslands of North America, Europe and Asia. These areas are sparsely populated with large farms spreading over hundreds of hectares. Severe winters restrict the growing season and only a single crop can be grown.

- ✰ In mixed farming the land is used for growing food and fodder crops and rearing livestock.
- ✰ It is practised in Europe, eastern USA, Argentina, southeast Australia, New Zealand and South Africa.
- ✰ Plantations are a type of commercial farming where single crop of tea, coffee, sugarcane, cashew, rubber, banana or cotton are grown. Large amount of labour and capital are required. The produce may be processed on the farm itself or in nearby factories. The development of a transport network is thus essential for such farming.
- ✰ Major plantations are found in the tropical regions of the world. Rubber in Malaysia, coffee in Brazil, tea in India and Sri Lanka are some examples.

Major Crops

A large variety of crops are grown to meet the requirement of the growing population. Crops also supply raw materials for agro-based industries. Major food crops are wheat, rice, maize and millets. Jute and cotton are fibre crops. Important beverage crops are tea and coffee.

- ✰ *Rice:* Rice is the major food crop of the world. It is the staple diet of the tropical and sub-tropical regions. Rice needs high temperature, high humidity and rainfall. It grows best in alluvial clayey soil, which can retain water. China leads in the production of rice followed by India, Japan, Sri Lanka and Egypt. In favourable climatic conditions as in West Bengal and Bangladesh two to three crops a year are grown.
- ✰ *Wheat:* Wheat requires moderate temperature and rainfall during growing season and bright sunshine at the time of harvest. It thrives best in well drained loamy soil. Wheat is grown extensively in USA, Canada, Argentina, Russia, Ukraine, Australia and India. In India it is grown in winter.
- ✰ *Millets:* They are also known as coarse grains and can be grown on less fertile and sandy soils. It is a hardy crop that needs low rainfall and high to moderate temperature and adequate rainfall. Jowar, bajra and ragi are grown in India. Other countries are Nigeria, China and Niger.
- ✰ *Maize:* Maize requires moderate temperature, rainfall and lots of sunshine. It needs well-drained fertile soils. Maize is grown in North America, Brazil, China, Russia, Canada, India, and Mexico.

- ☆ *Cotton:* Cotton requires high temperature, light rainfall, two hundred and ten frost-free days and bright sunshine for its growth. It grows best on black and alluvial soils. China, USA, India, Pakistan, Brazil and Egypt are the leading producers of cotton. It is one of the main raw materials for the cotton textile industry.
- ☆ *Jute:* Jute was also known as the 'Golden Fibre'. It grows well on alluvial soil and requires high temperature, heavy rainfall and humid climate. This crop is grown in the tropical areas. India and Bangladesh are the leading producers of jute.
- ☆ *Coffee:* Coffee requires warm and wet climate and well drained loamy soil. Hill slopes are more suitable for growth of this crop. Brazil is the leading producer followed by Columbia and India.
- ☆ *Tea:* Tea is a beverage crop grown on plantations. This requires cool climate and well distributed high rainfall throughout the year for the growth of its tender leaves.
- ☆ It needs well-drained loamy soils and gentle slopes. Labour in large number is required to pick the leaves. Kenya, India, China, Sri Lanka produce the best quality tea in the world.

Chapter 11

History of Rice, Sugarcane and Cotton

History of Rice

An Overview

- ☆ Rice is a member of a family of plants that also includes marijuana, grass and bamboo. There are over 120,000 different varieties of rice including black and red strains as well as white ones. Rice plants can grow to a height of ten feet and shoot up as much as eight inches in a single day. Rice is the world's No.1 the world's most important food crop and dietary staple, ahead of wheat, corn and bananas.
- ☆ Rice grows almost anywhere: the flooded plains of Bangladesh, the terraced countryside of northern Japan, the Himalayan foothills of Nepal and even the deserts of Egypt and Australia. Rice straw was traditionally used make sandals, hats, ropes and patches for thatch roofs.

Most new strains of rice have been developed at the International Rice Research Institute (IRRI) in Los Banos, the Philippines. Founded in 1967, the IRRI has produced 300 different varieties of rice. One strain, called IR36, is resistant to 15 insect, disease and environmental stresses and is currently planted on about 27 million acres worldwide.

The IRRI Genetics Research Center houses samples from over 80,000 varieties of rice, which are cross bred to produce new strains. Strains with desirable characteristics are saved and breed again. Usually, it takes at least six generations for these characteristics to actually become a reproducible trait of the plant.

History

- ☆ It is believed to have been rice first cultivated in China or possibly somewhere else in eastern Asia around 10,000 years ago. The earliest concrete evidence of rice farming comes from a 7000-year-old archeological site near the lower Yangtze River village of Hemudu in Zheijiang province in China when the rice grains unearthed there were found they were white but exposure to air turned them black in a matter minutes. These grains can now be seen at a museum in Hemudu.
- ☆ According to a Chinese legend rice came to China tied to a dogs tail, rescuing people from a famine that occurred after a severe flood. Evidence of rice dated to 7000 B.C. has been found near the village of Jiahu in Henan Province northern China near the Yellow River. It is not clear whether the rice was cultivated or simply collected Rice gains dated to 6000 B.C. have been discovered Changsa in the Hunan Province. In the early 2000s, a team form South Korea's Chungbuk National University announced that it had found the remains of rice grains in the Paleolithic site of Sorori dated to around 12,000 B.C.
- ☆ The earliest evidence of rice farming in Japan was dated to around 300 B.C. which worked nicely into models that it was introduced when the Koreans, forced to migrate by upheaval in China n the Warring States Period (403-221 B.C.), arrived around the same time. Later a number of Korean objects, dated between 800 and 600 B.C., were found.
- ☆ In the early 2000s, grains of wetland rice were found in pottery from northern Kyushu dated to 1000 B.C. This called into question the dating of the entire Yayoi period and caused some archeologist to speculate that maybe wet-land rice farming was introduced directly from China. This assertion is backed up somewhat by similarity in skeletal remains of 3000-year-old skeletons found in Quinghai province in China and Yayoi bodies unearthed in northern Kyushu and Yamaguchi prefecture.
- ☆ Wild rice grows in forest clearings but was adapted to grow in shallow flooded fields. The introduction of paddy agriculture dramatically changed the landscape and ecology of the regions.
- ☆ DNA analysis shows that these early forms of rice were different from varieties eaten today. Africans cultivated another species of rice around 1500 B.C. The Moors introduced rice to Europe via Spain.

Early Rice Farming in Japan

- ☆ As many archeologist looked upon the introduction of wet land rice farming techniques as the technological advancement that marked the beginning of the Yayoi period (400 B.C."A.D. 300) and the end of the Jomon period. In Kyushu people at red-kerneled rice.

- For a long time the earliest evidence of rice farming was dated to around 300 B.C. which worked nicely into models that it was introduced when the Koreans, forced to migrate by upheaval in China n the Warring States Period (403-221 B.C.), arrived around the same time.
- In the early 2000s, grains of wetland rice were found in pottery from northern Kyushu dated to 1000 B.C. This called into question the dating of the entire Yayoi period and caused some archeologist to speculate that maybe wet-land rice farming was introduced directly from China. This assertion is backed up somewhat by similarity in skeletal remains of 3000-year-old skeletons found in Quinghai province in China and Yayoi bodies unearthed in northern Kyushu and Yamaguchi prefecture. Japanese immigrants introduced superior rice varieties to the United States. By the early 1920s, 85 per cent of the rice in California was of Japanese origin.

Rice as Food

- The seeds in rice are contained in branching heads called panicles. Rice seeds, or grains, are 80 per cent starch. The remainder is mostly water and small amounts of phosphorus, potassium, calcium and B vitamins. Freshly harvested rice grains include a kernel made of an embryo (the heart of the seed), the endosperm that nourishes the embryo, a hull and several layers of bran which surround kernel. White rice consumed by most people is made up exclusively of kernels. Brown rice is rice that retains a few nutritious layers of bran.
- The bran and hull are removed in the milling process. In most places this residue is fed to livestock, but in Japan the bran is made into salad and cooking oil believed to prolong life. In Egypt and India it is made into soap. Eating unpolished rice prevents beriberi. The texture of rice is determined by a component in the starch called amylose. If the amylose content is low (10 to 18 per cent) the rice is soft and slightly sticky. If it is high (25 to 30 per cent) the rice is harder and fluffy. Chinese, Koreans and Japanese prefer their rice on the sticky side. People in India, Bangladesh and Pakistan like theirs fluffy, while people in Southeast Asia, Indonesia, Europe and the United States like theirs in between. Laotians like their rice gluey (2 per cent amylose).

Rice as a Food Crop

About 97 per cent of the world's rice is eaten within the country in which it is grown and most of this is cultivated with three miles of the people that eat it. About 92 per cent of the world crop is raised and consumed in Asia–a third in China and a fifth in India. Where irrigated paddy rice is grown one can find the densest populations. Rice supports 770 people per square kilometer in Yangtze and Yellow river basins in China and 310 per square kilometer in Java and Bangladesh.

Over 520 million tons of rice is harvested every year and about one tenth of all cultivated acreage in the world is devoted to rice. More corn and wheat is produced than rice but over 20 per cent of all wheat and 65 per cent of all corn is used for feeding livestock. Almost all rice is eaten by people not animals.

The Balinese eat about a pound of rice a day. The Burmese consume a little more than a pound; Thais and Vietnamese about three quarters of a pound; and the Japanese about a third of a pound. In contrast, the average America eats about 22 pounds a year. A tenth of rice grown in the United States is used in making beer. It provides a "lighter color and more refreshing taste," an Anheuser-Busch brewmaster told National Geographic.

Rice Planting

- ☆ Rice is one of the world's most labor intensive foods. In Japan the planting and harvesting is done mostly with machines, but in much of the world these chores–along with weeding, and maintaining the paddies and irrigation canals–are still largely done by hand, with water buffalo helping with the plowing and preparing of the fields. Traditionally rice has been harvested with a scythe, left to dry on the ground for a couple of days, and bundled into sheaves. Between 1000 to 2000 man or women hours are required to raise a crop on 2.5 acres of land.
- ☆ The fact that rice is so labor intensive tends to keep a lot of the population on the land. Rice is also a eater thirsty crop requiring lots of rain or irrigation water the wet rice grown in most Asia, needs hot weather after a period of rain conditions provided by the monsoons that affected many of the places where rice is grown.
- ☆ Fields are prepared before the rainy season with some plowing, often using water buffalo, and flooding. About a week or before planting so the paddy is partially drained, leaving behind a thick, muddy soup. Rice seedlings are grown in nursery plots, transplanted by hand or with a machine. Seedling are planted instead of seeds because the young plants are less vulnerable to disease and weeds than the seeds. Farmers that can afford pesticides and fertilizers sometimes plant seeds.
- ☆ Rice planting in much of the world is still done by hand, using methods that for the most part have remained unchanged for the last three of four thousand years. The foot-long seedlings are planted a couple at a time by bent-over planters who use their thumb and middle fingers to push the seedlings in the mud.
- ☆ Good planters average about one insertion a second in a process that the travel writer Paul Theroux once said was more like needlepoint than farming. The sticky, black mud in the paddy is usually ankle deep, but sometime knee deep, and rice planter generally go barefoot instead of wearing boots because the mud sucks the boots off.

- ☆ In Japan, Korea and other countries, farmers now use small diesel-powered rototiller-tractors to plow the rice paddies and refrigerator-size mechanical rice transplanters to plant the rice seedlings. In the old days it took 25 to 30 people to transplant the seedlings of one rice paddy. Now a single mechanical rice transplanter can do the job in a couple dozen paddies in one day.
- ☆ The seedling come on perforated plastic trays, which are placed directly on the transplanter. which uses a hook-like device to pluck the seedlings from the trays and plant them in the ground.

Growing and Harvesting

- ☆ The water depth in the paddy is increased as the rice seedlings grow and then gradually lowered in increments until the field is dry when the rice is ready to be harvested. Sometimes the water is drained during the growing season so the field can be weeded and the soil aerated and then water is put back in.
- ☆ Rice is harvested when it is a golden-yellow color several weeks after water has been completely drained from the paddy and the soil around the rice is dry. In many places rice is still harvested with a sickle and bundled into sheaves and then threshed by cutting the top inch or so of the stalks with a knife and removing the grains by slapping the stalks over propped up boards. The rice is placed on large sheets and left to dry on the ground for a couple of days before being taken to the mill to be processed. In many villages around the world, farmers usually help each other to harvest their crops.
- ☆ There are also harvesting machines. Some diesel-powered rototiller-tractors and mechanical rice transplanters are available with harvesting attachments. Large machines are not used to harvest rice because they can not maneuver around the paddy without messing them up. Plus, most rice paddies are small and divided by dikes. Large machines need long tracts of uniform land to do their job efficiently.
- ☆ After the rice harvest the stubble is often burned along with waste products from the harvest and the ashes are plowed back into the field to fertilize it.

Ecological Advantages

- ☆ Unlike slash and burn agriculture, which can sustainably support only 130 people per square mile, often seriously damaging the soil and filling the air with smoke, rice cultivation can support 1,000 people and not deplete the soil."
- ☆ Rice is unique as a crop in that it can grow in flooded conditions that would drown other plants (some rice species grow in water 16 feet deep).

What makes this possible is an efficient air-gathering system consisting of passages in the upper leaves of rice plants that draw in enough oxygen and carbon dioxide to nourish the entire plant.

- ☆ Nitrogen is the most important plant nutrient and fortunately for rice growers blue-green algae, one of two organism on earth that can transform oxygen from the air into nitrogen, thrives in the stagnant rice paddy water. The decayed algae as well as old rice stalks and other decomposed plants and animals provide nearly all the nutrients for growing rice plants, plus they leave behind enough nutrients for future crops."
- ☆ The constant supply of nutrients means that the paddy soils are resilient and don't become worn out like other soils. In flooded rice paddies few nutrients are leached (carried away by rain water deep into the soil where plants can't get them) and the nutrients dissolved in the murky water are easy for the plant to absorb. In tropical climates two, sometimes three, rice crops can be raised every year."
- ☆ Rice paddies create a lovely landscape and have their own rich ecosystem. Fish such as minnows, loaches and bitterling can survive in the paddies and the canals as can aquatic snails, worms, frogs, crawfish beetles, fireflies and other insects and even some crabs. Egrets, kingfishers, snakes and other birds and predators feed on feed on these creatures. Ducks have been brought into rice paddies to eat weeds and insects and eliminate the need for herbicides and pesticides. Innovations such as concrete-sided canals have damaged the rice paddy ecosystem by depriving plants and animals of places they can live.

Major Problems

Major problems in rice cultivation are summarised as under:

- ☆ Rice grains Bacterial leaf blight, plant hoppers, rodents and stem borders are the major rice destroying pests. These days the biggest threat to the world rice crops is the leaf blight, a disease which wipes out as much as half the rice crop in some parts of Africa and Asia, and annually destroys between 5 and 10 per cent of the world's total rice harvest. In 1995, scientist cloned a gene that protects rice plants from leaf blight and developed a genetically-engineered and cloned rice plant that resists the disease.
- ☆ The trend towards reliance on only a few strains of highly-productive rice plants worldwide has the potential of causing a disaster. If these strains suddenly become vulnerable to a disease or pests, huge amounts of crops could be destroyed, causing severe food shortages or even a famine. If many strains are used and some of them are destroyed by disease or pests, there are still many remaining stains producing rice and overall food supply is not jeopardized.

- While demands for food increase, land used to grow rice is being lost to urbanization and industry and the demands of a growing population. Demographers estimate that production of rice must increase by 70 per cent over the next 30 years to keep up with a population that is supposed to grow by 58 per cent before the year 2025.
- Much of the rice grown in coastal plains and river deltas is vulnerable to sea-level rises caused by global warming. Sometimes fertilizers and pesticides leak out of the paddies and damage the environment.

Processing

- After rice is harvested it is usually taken to a local rice processing plant where huge two-story machines with conveyor belts and vibrators removes the husks and other impurities, polish the rice and deposit in a bag ready to eat. The best machines come from Japan. The husks and other left over materials are fed to pigs or other animals. Milling 10 kilograms of rice with a machine takes about 5 minutes. By hand it takes it two to three hours.
- Describing the "modern way" of husking rice used in some parts of Asia, Patrick Tyler wrote in the New York Times, "First a tractor was driven, in reverse, in fast and furious circles over the dried rice stalks. The process resembled a dangerous rodeo stunt as the tractor bounded over 5- and 6-foot piles of rice stalks, narrowly missing farm hands, women and children who pushed and pulled at the straw to insure that all of it got run over several times."
- "The crazy dance of the tractor caused most of the rice grains still inside the husks to separate from the stalks," Tyler wrote. "The peasants then bundled and beat the stalks against the side of a wooden box, collecting as many grains as they could. The stalks were then built into stacks, which would be burned to fertilize the fields.The rice grains were left on the concrete slab mixed with the debris of the stalks. After a few more hours of raking and hurling the debris into the air, the rice was finally separated from the chaff and the harvest was bagged for storage."

Improved Rice Yields

- Rice plants have been greatly improved over the last three decades. Plant breeders have created plants that mature in 110 days instead 160 which means that regions with warm climates can grow three crops instead of two. The height of the average plant has been reduced from five feet to a stocky three feet, which means that the plant nutrient go into producing grains of rice and are not "wasted" on the stalks that lean over when there is too much weight. In addition, rice plants have been bred and bio-engineered to be resistant to bacterial blight, plant hoppers and stem borders.

- ☆ Tall conventional rice plant used before 1968 grew in 140-180 days and yielded between 0.6 and 1.4 tons per acre. Modern rice grows in 110-140 days, produces 100 seeds per panicle, and yields between 2.4 and 4.0 tons per acre.
- ☆ The introduction of new more productive strains of rice (and corn and wheat) to developing countries like India and Indonesia in the 1960s was called the "Green Revolution." Before the Green revolution India had problems feeding itself, now it exports rice. Between 1967 and 1992 the world's rice harvest has doubled. in Indonesia it tripled from 15 million tons to 48 million tons.
- ☆ Current dwarf varieties have 15 productive panicles, or seed clusters per stalk (out 25 or so total stalks) that produce about 100 grains (seeds) each. New strains will have fewer, but stronger and thicker, stalks that will yield 200 or more grains each. These new plants are expected to account for most of the increased productivity.

Super Rice and Golden Rice

- ☆ Scientists at the IRRI in 1995 announced that they had developed a "super rice" that grows in 100-130 days, yields up to 5.3 tons per acre, and produces 200 to 250 seeds per panicle, compared to 100 in most modern rice varieties. The goal of the super rice program was to produce a plant that devotes its energy to produce grain in panicles rather leaves and stems.
- ☆ When it is available the new rice is expected to increase yields in irrigated fields by 25 per cent (from 10 tons per hectare to 12 tons per hectare) with less fertilizer than modern varieties. It is hoped that the "super rice" will hope to keep up with demands made by a growing population and offset the loss of agricultural to industry.
- ☆ Super rice will not be introduced until it has been crossbreed to produce disease-resistant strain that can grown in a variety of climates and conditions. In the late 1990s, super rice strains were given to rice breeders in different parts of the world to be crossbred with varieties suited to local soils, temperatures and rainfall and could deal with a pests and diseases it is expected to encounter.
- ☆ Golden rice is a kind of rice that has been genetically engineered to contain beta-carotene, a substance gives carrots their orange color and helps the human body can convert to Vitamin A Golden rice has been modified with a daffodil gene to produce beta-carotene, which the normally white grains a pale yellow color.
- ☆ Golden rice has been marketed as a way to reduce blindness caused by Vitamin A deficiency among children in the Third World, where an estimated 500,000 people go blind from Vitamin A deficiency every

year. A closer look at golden rice however shows that it is not all that it is cranked up to. A child would have to eat 15 pounds of golden rice a day to get the daily minimum requirement of Vitamin A.

History of Sugarcane

An Overview

History of the sugar and the world economy would look drastically different without the presence of sugarcane, incredible plant that drove many changes in our modern history and forged the basis of the modern cuisine. With its incredible ability to create and store sucrose in large quantities, sugarcane went from unknown wild species of Asian perennial true grasses to the world's largest cultivated crop.

- Sugarcane is a tropical, perennial grass that forms lateral shoots at the base to produce multiple stems, typically three to four m (10 to 13 ft) high and about 5 cm (2 in) in diameter. The stems grow into cane stalk, which when mature constitutes around 75 per cent of the entire plant. A mature stalk is typically composed of 11–16 per cent fiber, 12–16 per cent soluble sugars, 2–3 per cent non-sugars, and 63–73 per cent water.
- A sugarcane crop is sensitive to the climate, soil type, irrigation, fertilizers, insects, disease control, varieties, and the harvest period. The average yield of cane stalk is 60–70 tonnes per hectare (24–28 long ton/acre; 27–31 short ton/acre) per year. However, this figure can vary between 30 and 180 tonnes per hectare depending on knowledge and crop management approach used in sugarcane cultivation. Sugarcane is a cash crop, but it is also used as livestock fodder.
- Sugarcane is a grass plant of the genus *Saccharum*, tribe Andropogoneae that can be found in 36 species variety. It was originally native to warm tropical regions of Asia, but after early civilizations found out about its usefulness it quickly spread. This enabled new civilizations to improve their sugar production with crossbreeding (all current commercial sugarcanes are complex hybrids), which only increased sugarcane's popularity.
- Sugarcane was originally domesticated around 8000 BC in New Guinea. From there knowledge about this plant slowly moved toward east across Southeast Asia until it reached India, where the first organized production of sugar began during middle of 1st millennia BC. In the beginning sugar was extracted from sugarcanes by chewing and extracting fluids via water, but in 5th century AD Indian chemist found a way to crystallize extracted sucrose, making sugar much easier to transport. With this great discovery, sugar became very expensive trading item of India, and sugarcanes themselves started spreading across the Asia and Middle East.

- Arab nations adopted sugarcanes, and spread them toward Mediterranean, especially after they managed to conquer Egypt. From there sugarcanes reached Spain around 715 AD, but they did not took hold until Crusades, when majority of the Europe became acquainted with the sugar that was imported from the Middle East. Spain, Portugal, Italy, Cypress and Azores tried to establish stable economy around sugarcanes in 16th and 17th century, but vast new lands in the New World offered better climate. Because of that, sugarcanes were quickly introduced to the Americas, where landowners created vast plant ages of this plant. Because growing and processing of sugarcanes was not easy, and demand for sugar was extremely high, organized culling of Africa for slaves became very profitable and popular. With over 12 million slaves displaced from Africa, production of sugar skyrocketed, managing to drop in price and become available to everyone during 19th century.
- Today, sugarcane is the world largest crop. In 2010 it was estimated that over 23.8 million hectares of sugarcanes were cultivated in over 90 countries around the world, with a worldwide harvest of 1.69 billion tonnes. The largest producer of Sugarcanes is Brazil, and behind him are India, China, Pakistan, Thailand and Mexico. Also, sugarcane represents a source for 75-80 per cent of worldwide sugar production, with the majority of the rest being taken by sugar beet that is more suited to grow in Europe.

- Sugarcane is indigenous to tropical parts of South and Southeast Asia. Different species likely originated in different locations, with *Saccharum barberi* originating in India and *S. edule* and *S. officinarum* in New Guinea. The earliest known production of crystalline sugar began in northern India. The exact date of the first cane sugar production is unclear. The earliest evidence of sugar production comes from ancient Sanskrit and Pali texts.
- In the 8th century, Muslim and Arab traders introduced sugar from South Asia to the other parts of the Abbasid Caliphate in the Mediterranean, Mesopotamia, Egypt, North Africa, and Andalusia. By the 10th century, sources state that every village in Mesopotamia grew sugarcane. It was among the early crops brought to the Americas by the Spanish, mainly Andalusians, from their fields in the Canary Islands, and the Portuguese from their fields in the Madeira Islands.
- Christopher Columbus first brought sugarcane to the Caribbean during his second voyage to the Americas; initially to the island of Hispaniola (modern day Haiti and the Dominican Republic). In colonial times, sugar formed one side of the triangle trade of New World raw materials, along with European manufactured goods, and African slaves. Sugar (often in the form of molasses) was shipped from the Caribbean to Europe or New

England, where it was used to make rum. The profits from the sale of sugar were then used to purchase manufactured goods, which were then shipped to West Africa, where they were bartered for slaves. The slaves were then brought back to the Caribbean to be sold to sugar planters. The profits from the sale of the slaves were then used to buy more sugar, which was shipped to Europe.

- France found its sugarcane islands so valuable that it effectively traded its portion of Canada, famously dubbed "a few acres of snow", to Britain for their return of Guadeloupe, Martinique and St. Lucia at the end of the Seven Years' War. The Dutch similarly kept Suriname, a sugar colony in South America, instead of seeking the return of the New Netherlands (New York).
- Boiling houses in the 17th through 19th centuries converted sugarcane juice into raw sugar. These houses were attached to sugar plantations in the Western colonies. Slaves often ran the boiling process under very poor conditions. Rectangular boxes of brick or stone served as furnaces, with an opening at the bottom to stoke the fire and remove ashes. At the top of each furnace were up to seven copper kettles or boilers, each one smaller and hotter than the previous one. The cane juice began in the largest kettle. The juice was then heated and lime added to remove impurities. The juice was skimmed and then channeled to successively smaller kettles. The last kettle, the "teache", was where the cane juice became syrup. The next step was a cooling trough, where the sugar crystals hardened around a sticky core of molasses. This raw sugar was then shoveled from the cooling trough into hogsheads (wooden barrels), and from there into the curing house.
- In the British Empire, slaves were liberated after 1833 and many would no longer work on sugarcane plantations when they had a choice. British owners of sugarcane plantations therefore needed new workers, and they found cheap labour in China, Portugal and India. The people were subject to indenture, a long-established form of contract which bound them to forced labour for a fixed term; apart from the fixed term of servitude, this resembled slavery.[13] The first ships carrying indentured labourers from India left in 1836. The migrations to serve sugarcane plantations led to a significant number of ethnic Indians, southeast Asians and Chinese settling in various parts of the world. In some islands and countries, the South Asian migrants now constitute between 10 and 50 per cent of the population. Sugarcane plantations and Asian ethnic groups continue to thrive in countries such as Fiji, Natal, Burma, Sri Lanka, Malaysia, Indonesia, Philippines, British Guiana, Jamaica, Trinidad, Martinique, French Guiana, Guadeloupe, Grenada, St.Lucia, St. Vincent, St.Kitts, St.Croix, Suriname, Nevis, and Mauritius.
- Between 1863 and 1900, traders and plantation owners from the British

colony of Queensland (now a state of Australia) brought between 55,000 and 62,500 people from the South Pacific Islands to work on sugarcane plantations. It is estimated that one-third of these workers were coerced or kidnapped into slavery (known as blackbirding). Many others were paid very low wages. Between 1904 and 1908, most of the 10,000 remaining workers were deported in an effort to keep Australia racially pure and protect white workers from cheap foreign labour.

- ✰ Cuban sugar derived from sugarcane was exported to the USSR, where it received price supports and was ensured a guaranteed market. The 1991 dissolution of the Soviet state forced the closure of most of Cuba's sugar industry.
- ✰ Sugarcane remains an important part of the economy of Guyana, Belize, Barbados, and Haiti, along with the Dominican Republic, Guadeloupe, Jamaica, and other islands. About 70 per cent of the sugar produced globally comes from *S. officinarum* and hybrids using this species.

Cultivation

- ✰ Sugarcane cultivation requires a tropical or subtropical climate, with a minimum of 60 cm (24 in) of annual moisture. It is one of the most efficient photosynthesizers in the plant kingdom. It is a C4 plant, able to convert up to 1 per cent of incident solar energy into biomass. In prime growing regions, such as Mauritius, Dominican Republic, Puerto Rico, Peru, Brazil, Bolivia, Colombia, Guyana, Ecuador, Cuba, El Salvador, Jamaica, Bangladesh, India, Pakistan, Indonesia, Philippines, Malaysia, Australia and Hawaii, sugarcane crops can produce over 15 kg/m^{2}of cane. Once a major crop of the southeastern region of the United States, sugarcane cultivation has declined there in recent decades, and is now primarily confined to Florida, Louisiana, and South Texas.
- ✰ Sugarcane is cultivated in the tropics and subtropics in areas with a plentiful supply of water for a continuous period of more than six to seven months each year, either from natural rainfall or through irrigation. The crop does not tolerate severe frosts. Therefore, most of the world's sugarcane is grown between 22°N and 22°S, and some up to 33°N and 33°S.When sugarcane crop is found outside this range, such as the Natal region of South Africa, it is normally due to anomalous climatic conditions in the region, such as warm ocean currents that sweep down the coast. In terms of altitude, sugarcane crop is found up to 1,600 metres or 5,200 feet close to the equator in countries such as Colombia, Ecuador, and Peru.
- ✰ Sugarcane can be grown on many soils ranging from highly fertile well-drained mollisols, through heavy cracking vertisols, infertile acid oxisols, peaty histosols, to rocky andisols. Both plentiful sunshine and water supplies increase cane production. This has made desert countries with

good irrigation facilities such as Egypt some of the highest-yielding sugarcane-cultivating regions.

- Some sugarcanes produce seeds, modern stem cutting has become the most common reproduction method. Each cutting must contain at least one bud, and the cuttings are sometimes hand-planted. In more technologically advanced countries like the United States and Australia, billet planting is common. Billets (stalks or stalk sections) harvested by a mechanical harvester are planted by a machine that opens and recloses the ground. Once planted, a stand can be harvested several times; after each harvest, the cane sends up new stalks, called ratoons. Successive harvests give decreasing yields, eventually justifying replanting. Two to 10 harvests are usually made depending on the type of culture. In a country with a mechanical agriculture looking for a high production of large fields, like in North America, sugar canes are replanted after two or three harvests to avoid a lowering in yields. In countries with a more traditional type of agriculture with smaller fields and hand harvesting, like in the French island la Réunion, sugar cane is often harvested up to 10 years before replanting.
- Sugarcane is harvested by hand and mechanically. Hand harvesting accounts for more than half of production, and is dominant in the developing world. In hand harvesting, the field is first set on fire. The fire burns dry leaves, and chases away or kills any lurking venomous snakes, without harming the stalks and roots. Harvesters then cut the cane just above ground-level using cane knives or machetes. A skilled harvester can cut 500 kg (1,100 lb) of sugarcane per hour.
- Mechanical harvesting uses a combine, or sugarcane harvester. The Austoft 7000 series, the original modern harvester design, has now been copied by other companies, including Cameco/John Deere. The machine cuts the cane at the base of the stalk, strips the leaves, chops the cane into consistent lengths and deposits it into a transporter following alongside. The harvester then blows the trash back onto the field. Such machines can harvest 100 long tons (100 t) each hour; however, harvested cane must be rapidly processed. Once cut, sugarcane begins to lose its sugar content, and damage to the cane during mechanical harvesting accelerates this decline. This decline is offset because a modern chopper harvester can complete the harvest faster and more efficiently than hand cutting and loading. Austoft also developed a series of hydraulic high-lift infield transporters to work alongside their harvesters to allow even more rapid transfer of cane to, for example, the nearest railway siding. This mechanical harvesting doesn't require the field to be set on fire; the remains left in the field by the machine consist of the top of the sugar cane and the dead leaves, which act as mulch for the next round of planting.

History of Cotton

An Overview

- ☆ Cotton is arguably the world's most important nonfood crop. It supplies about 80 per cent of the world's natural fibers and continues to be primary material in half the world's textiles despite competition from synthetic fibers. Cotton is ideal for making cloth because its fibers bond and interlock when spun into long strands that can be is easily dyed. *[Source: Jon Thompson, National Geographic, June 1994]*

Cotton accounts for 40 per cent of the fiber used today. About 20 million farmers in 80 countries produce 44 million metric tons of cotton a year. Cotton was once called white gold. Developing the machinery to quickly make it into cloth was a driving force behind the Industrial Revolution. Providing labour to produce it for these machine was behind the slave economy in the southern United States and in turn and indirect cause of the American Civil War. Making money from it kept the colonial economies going in places like India and Egypt.

- ☆ Cotton is used to make clothes, towels, sheets, rugs, draperies, cloth, batting, cellulose products, cordage, and sewing threads. It can also be ground into currency, crushed into vegetable oil, and woven into coffee filters, book bindings, tents, diapers and fishnet. Fire men like cotton firehouses over polyester ones because they soak up enough water to keep them from melting. Environmentalists like it mops up oil spills better than almost any other material. And doctors like for bandages and sutures because of its durability under a lot of different conditions.

Weaving and Cloth

- ☆ Weaving is the interlocking at right angles of two sets of fibers to make cloth or a similar material. Spinning is the process by which fibers are drawn and twisted into string, yarn or thread. These tasks have traditionally been done at home by women with looms and spinning devices. The advantage with working at home is that women could work and care for their children at the same time and do weaving and spinning when they are not doing other chores.
- ☆ To make yarn the old-fashion way:
  1) fibers are cleaned and straightened by carding with a toothed card (a sort of brush with stiff bristles).
  2) The fibers are then rolled between two cards to produce a thin fiber-like piece called a sliver.
  3) The sliver is placed on a spike called a distaff.

4) A strand of wool is then pulled off and a weight known as whorl is attached to it.
5) The strand is twisted into a thread by spinning it with the thumb and forefinger. Since each thread is made this way, you can how time consuming it must have been to make a piece of cloth before machinery was widely used.

☆ To make cloth the old-fashion way:

1) Threads are placed on a warp-weighted loom. Warps are the downward hanging threads on a loom, and they are set up so that every other thread faces forward and the others are in the back.
2) A weft (horizontal thread) is then taken in between the forward and backward row of warps.
3) Before the weft is threaded through in the other direction, the position of the warps is changed with something called a heddle rod.

   This simple tool reverses the warps so that the row in the front is now in the rear, and visa versa. In this way the threads are woven in a cross stitch manner that holds them together and creates cloth. The cloth in turn is all kinds of things. Cloth is often sold to clothing manufacturers in rolls. One roll of cloth is about 11 meters in length.

☆ Sometimes natural dies are used. Red is made with madder or kermes (an insect found in kermes oak); yellow with wild chamomile, saffron, vine leaves and the rinds of pomegranates; black from acorns; and blue from plants with indigo. Natural dyes produce richer and more natural colors. The depth of color can be controlled by the type of water–rain, river or spring–used. Plant dyes though are notoriously unpredictable. They are affected by weather conditions, soil and when they are harvested. They have mostly been replaced by chemical dyes.

Industrial Revolution

☆ The Industrial Revolution came about as an effort to devise faster ways to make cotton cloth and print it. One of the most important technological advances of that period was the water-powered spinning machine, which efficiently produced cotton thread for the first time. It was invented by Sir Richard Arkwright, a former barber who used fashion wigs from the hair he clipped from his clients, in the Derbyshire village of Cromford in 1771 and immediately put to work in a factory opened by Arkwright.

☆ In During the 18th century English women had taken a fancy to comfortable garments made from cloth woven in India. Arkwright and others realized that large profits could be made by growing cotton in the America and spinning it into thread and cloth in England. What is now the southern United States as selected because they were closer to Britain than India and had ideal climate for growing the crop.

- In the year 1793 the cotton industry was given another boost when Eli Whitney invented the cotton gin, a hand cranked devise with a roller with teeth that tore the cotton into bits and removed the fluffy fibers (or lint) from seeds. By the mid 1800's America was exporting nearly two billion pounds of cotton a year. Most of it was being shipped into Liverpool and then transported by river and canal to Manchester for manufacturing. India did not become a major producer of cotton again until after the American Civil War stopped the supply of cotton to Britain from the United States.

Cotton, Colonialism and India

- Cotton grown in India was shipped to Manchester, England where it was made into finished goods which were sold back to India for a tidy profit. Gandhi began his home spinning movement and making of clothes to foil this trade. The British exploited Egypt in a similar way. Cotton starting flowing out of India after the supply from the United States was cut off by the American Civil War. Cotton originated from South Asia.
- Cotton undergarments made in India in the 16th century were said be so delicate and beautifully embroidered they only lasted for a couple of hours. It was no surprise that Gandhi's first fast in 1918 was conducted to support textile workers in Ahmedabad striking for higher wages. The factory in Ahmedabad where the men worked is still running today. "To drive the looms," wrote Jon Thompson in National Geographic,"huge wheels clattered and clanked, now run by electricity but powered by steam engines in the 19th century. Cotton dust had accumulated on the windows and on every pipe and loom and wire. Overhead pipes sprayed mist into the air to moisten the cotton fibre. The workers wore only loin clothes because of the unbearable heat and humidity. Although some of the workers were speaking, no sound came from their mouths.
  Source: [Jon Thompson, National Geographic, June 1994]
- Cotton played an important role in the history of India, the British Empire, and the United States, and continues to be an important crop and commodity. The history of the domestication of cotton is very complex and is not known exactly.
- Several isolated civilizations in both the Old and New World independently domesticated and converted cotton into fabric. All the same tools were invented, including combs, bows, hand spindles, and primitive looms

Early History

Americas

- Some of the oldest cotton bolls were discovered in a cave in Tehuacán Valley, Mexico, and were dated to approximately 5500 BCE, but some doubt has been cast on these estimates. Seeds and cordage dating to about 2500 BCE have been found in Peru. By 3000 BCE cotton was being grown and processed in Mexico, and Arizona.

- Pre-Incan cotton grave cloths were found in Huaca Prieta in Peru, and date back to 2500 BCE.

Indian Subcontinent

- The latest archaeological discovery in Mehrgarh puts the dating of early cotton cultivation and the use of cotton to 5000 BCE. The Indus Valley civilization started cultivating cotton by 3000 BCE.
- Cotton was mentioned in Hindu hymns in 1500 BCE. Herodotus, an ancient Greek historian, mentions Indian cotton in the 5th century BCE as "a wool exceeding in beauty and goodness that of sheep." When Alexander the Greatinvaded India, his troops started wearing cotton clothes that were more comfortable than their previous woolen ones.
- Strabo, another Greek historian, mentioned the vividness of Indian fabrics, and Arrian told of Indian–Arab trade of cotton fabrics in 130 CE.

Middle Ages

Eastern world

- Handheld roller cotton gins had been used in India since the 6th century, and was then introduced to other countries from there. Between the 12th and 14th centuries, dual-roller gins appeared in India and China. The Indian version of the dual-roller gin was prevalent throughout the Mediterranean cotton trade by the 16th century. This mechanical device was, in some areas, driven by water power.
- The spinning wheel was invented in India, between 500 and 1000 AD. The earliest clear illustrations of the spinning wheel come from the Islamic world in the eleventh century.

Western world

- Egyptians grew and spun cotton from 6–700 CE.
- Cotton was a common fabric during the Middle Ages, and was hand-woven on a loom. Cotton manufacture was introduced to Europe during the Muslim conquest of the Iberian Peninsula and Sicily. The knowledge of cotton weaving was spread to northern Italy in the 12th century, when Sicily was conquered by the Normans, and consequently to the rest of Europe. The spinning wheel, introduced to Europe circa 1350, improved the speed of cotton spinning. By the 15th century, Venice, Antwerp, and Haarlem were important ports for cotton trade, and the sale and transportation of cotton fabrics had become very profitable.

- Christopher Columbus, in his explorations of the Bahamas and Cuba, found natives wearing cotton ("the costliest and handsomest. cotton mantles and sleeveless shirts embroidered and painted in different designs and colours"), a fact that may have contributed to his incorrect belief that he had landed on the coast of India.

Early Modern Period

India

- India had been an exporter of fine cotton fabrics to other countries since the ancient times. Sources such as Marco Polo, who traveled India in the 13th century, Chinese travelers, who traveled Buddhist pilgrim centers earlier, Vasco Da Gama, who entered Calicut in 1498, and Tavernier, who visited India in the 17th century, have praised the superiority of Indian fabrics.
- The worm gear roller cotton gin, which was invented in India during the early Delhi Sultanate era of the 13th–14th centuries, came into use in the Mughal Empire some time around the 16th century, and is still used in India through to the present day. Another innovation, the incorporation of the crank handle in the cotton gin, first appeared in India some time during the late Delhi Sultanate or the early Mughal Empire.
- The production of cotton, which may have largely been spun in the villages and then taken to towns in the form of yarn to be woven into cloth textiles, was advanced by the diffusion of the spinning wheel across India shortly before the Mughal era, lowering the costs of yarn and helping to increase demand for cotton. The diffusion of the spinning wheel, and the incorporation of the worm gear and crank handle into the roller cotton gin, led to greatly expanded Indian cotton textile production during the Mughal era.
- It was reported that, with an Indian cotton gin, which is half machine and half tool, one man and one woman could clean 28 pounds of cotton per day. With a modified Forbes version, one man and a boy could produce 250 pounds per day. If oxen were used to power 16 of these machines, and a few people's labour was used to feed them, they could produce as much work as 750 people did formerly.
- During the early 16th century to the early 18th century, Indian cotton production increased, in terms of both raw cotton and cotton textiles. The Mughals introduced agrarian reforms such as a new revenue system that was biased in favour of higher value cash crops such as cotton and indigo, providing state incentives to grow cash crops, in addition to rising market demand.
- The largest manufacturing industry in the Mughal Empire was cotton textile manufacturing, which included the production of piece goods,

calicos, and muslins, available unbleached and in a variety of colours. The cotton textile industry was responsible for a large part of the empire's international trade. India had a 25 per cent share of the global textile trade in the early 18t^h century. Indian cotton textiles were the most important manufactured goods in world trade in the 18th century, consumed across the world from the Americas to Japan. The most important center of cotton production was the Bengal Subah province, particularly around its capital city of Dhaka.

- Bengal accounted for more than 50 per cent of textiles imported by the Dutch from Asia, Bengali cotton textiles were exported in large quantities to Europe, Indonesia, and Japan, and Bengali muslin textiles from Dhaka were sold in Central Asia, where they were known as "*daka*" textiles. Indian textiles dominated the Indian Ocean trade for centuries, were sold in the Atlantic Ocean trade, and had a 38 per cent share of the West African trade in the early 18th century, while Indian calicos were a major force in Europe, and Indian textiles accounted for 20 per cent of total English trade with Southern Europe in the early 18th century.

Western World

- Cotton cloth started to become highly sought-after for the European urban markets during the Renaissance and the Enlightenment. Vasco da Gama (d. 1524), a Portuguese explorer, opened Asian sea trade, which replaced caravans and allowed for heavier cargo. Indian craftspeople had long protected the secret of how to create colourful patterns. However, some converted to Christianity and their secret was revealed by a French Catholic priest, Father Coeurdoux (1691–1779). He revealed the process of creating the fabrics in France, which assisted the European textile industry.
- In early modern Europe, there was significant demand for cotton textiles from Mughal India. European fashion, for example, became increasingly dependent on Mughal Indian textiles. From the late 17th century to the early 18th century, Mughal India accounted for 95 per cent of British imports from Asia, and the Bengal Subah province alone accounted for 40 per cent of Dutch imports from Asia. In contrast, there was very little demand for European goods in Mughal India, which was largely self-sufficient, thus Europeans had very little to offer, except for some woolens, unprocessed metals and a few luxury items. The trade imbalance caused Europeans to export large quantities of gold and silver to Mughal India in order to pay for South Asian imports.

Egypt

- Egypt under Muhammad Ali in the early 19th century had the fifth most productive cotton industry in the world, in terms of the number of spindles per capita.

- ☆ The industry was initially driven by machinery that relied on traditional energy sources, such as animal power, water wheels, and windmills, which were also the principle energy sources in Western Europe up until around 1870. It was under Muhammad Ali of Egypt in the early 19th century that steam engines were introduced to the Egyptian cotton industry.

East India Company

- ☆ Cotton's rise to global importance came about as a result of the cultural transformation of Europe and Britain's trading empire. Calico and chintz, types of cotton fabrics, became popular in Europe, and by 1664 the East India Company was importing a quarter of a million pieces into Britain. By the 18th century, the middle class had become more concerned with cleanliness and fashion, and there was a demand for easily washable and colourful fabric.
- ☆ Wool continued to dominate the European markets, but cotton prints were introduced to Britain by the East India Company in the 1690s. Imports of calicoes, cheap cotton fabrics from Kozhikode, then known as *Calicut*, in India, found a mass market among the poor. By 1721 these calicoes threatened British manufacturers, and Parliament passed the Calico Act that banned calicoes for clothing or domestic purposes. In 1774 the act was repealed with the invention of machines that allowed for British manufacturers to compete with Eastern fabrics.
- ☆ Indian cotton textiles, particularly those from Bengal, continued to maintain a competitive advantage up until the 19th century. In order to compete with India, Britain invested in labour-saving technical progress, while implementing protectionist policies such as bans and tariffs to restrict Indian imports. At the same time, the East India Company's rule in India contributed to its deindustrialization, opening up a new market for British goods, while the capital amassed from Bengal after its 1757 conquest was used to invest in British industries such as textile manufacturing and greatly increase British wealth.
- ☆ British colonization also forced open the large Indian market to British goods, which could be sold in India without tariffs or duties, compared to local Indian producers who were heavily taxed, while raw cotton was imported from India without tariffs to British factories which manufactured textiles from Indian cotton, giving Britain a monopoly over India's large market and cotton resources. India served as both a significant supplier of raw goods to British manufacturers and a large captive market for British manufactured goods. Britain eventually surpassed India as the world's leading cotton textile manufacturer in the 19th century.
- ☆ The cotton industry grew under the British commercial empire. British cotton products were successful in European markets, constituting 40.5

per cent of exports in 1784–1786. Britain's success was also due to its trade with its own colonies, whose settlers maintained British identities, and thus, fashions. With the growth of the cotton industry, manufacturers had to find new sources of raw cotton, and cultivation was expanded to West India. High tariffs against Indian textile workshops, British power in India through the East India Company, and British restrictions on Indian cotton imports transformed India from the source of textiles to a source of raw cotton. Cultivation was also attempted in the Caribbean and West Africa, but these attempts failed due to bad weather and poor soil.

- The Indian subcontinent was looked to as a possible source of raw cotton, but intra-imperial conflicts and economic rivalries prevented the area from producing the necessary supply.

Britain

- Cotton's versatility allowed it to be combined with linen and be made into velvet. It was cheaper than silk and could be imprinted more easily than wool, allowing for patterned dresses for women. It became the standard fashion and, because of its price, was accessible to the general public. New inventions in the 1770s–such as the spinning jenny, the water frame, and the spinning mule–made the British Midlands into a very profitable manufacturing centre. In 1794–1796, British cotton goods accounted for 15.6 per cent of Britain's exports, and in 1804–1806 grew to 42.3 per cent.
- The Lancashire textile mills were major parts of the British industrial revolution. Their workers had poor working conditions: low wages, child labour, and 18-hour work days. Richard Arkwright created a textile empire by building a factory system powered by water, which was occasionally raided by the Luddites, weavers put out of business by the mechanization of textile production. In the 1790s, James Watt's steam power was applied to textile production, and by 1839; 200,000 children worked in Manchester's cotton mills. Karl Marx, who frequently visited Lancashire, may have been influenced by the conditions of workers in these mills in writing *Das Kapital.*

United States of America

Pre–Civil War

- Anglo-French warfare in the early 1790s restricted access to continental Europe, causing the United States to become an important–and temporarily the largest–consumer for British cotton goods. In 1791, U.S. cotton production was small, at only 900,000 kilograms. Several factors contributed to the growth of the cotton industry in the U.S. the increasing British demand; innovations in spinning, weaving, and steam power; inexpensive land; and a slave labour force. The modern cotton gin, invented in 1793 by Eli Whitney, enormously grew the American cotton industry, which was previously limited by the speed of manual removal of seeds from the fibre, and helped cotton to surpass tobacco as the primary cash crop of the South.

- By 1801 the annual production of cotton had reached over 22 million kilograms, and by the early 1830s the United States produced the majority of the world's cotton. Cotton also exceeded the value of all other United States exports combined. The need for fertile land conducive to its cultivation lead to the expansion of slavery in the United States and an early 19th-century land rush known as Alabama Fever.
- Cultivation of cotton using black slaves brought huge profits to the owners of large plantations, making them some of the wealthiest men in the U.S. prior to the Civil War. In the non-slave-owning states, farms rarely grew larger than what could be cultivated by one family due to scarcity of farm workers. In the slave states, owners of farms could buy many slaves and thus cultivate large areas of land. By the 1850s, slaves made up 50 per cent of the population of the main cotton states: Georgia, Alabama, Mississippi, and Louisiana.
- Slaves were the most important asset in cotton cultivation, and their sale brought profits to slaveowners outside of cotton-cultivating areas. Thus, the cotton industry contributed significantly to the Southern upper class's support of slavery. Although the Southern small-farm owners did not grow cotton due to its lack of short-term profitability, they were still supportive of the system in the hopes of one day owning slaves.
- Cotton's central place in the national economy and its international importance led Senator James Henry Hammond of South Carolina to make a famous boast in 1858.
- Without firing a gun, without drawing a sword, should they make war on us, we could bring the whole world to our feet. What would happen if no cotton was furnished for three years?. England would topple headlong and carry the whole civilized world with her save the South. No, you dare not to make war on cotton. No power on the earth dares to make war upon it. Cotton is king.
- Cotton diplomacy, the idea that cotton would cause Britain and France to intervene in the Civil War, was unsuccessful. It was thought that the Civil War caused the Lancashire Cotton Famine, a period between 1861–1865 of depression in the British cotton industry, by blocking off American raw cotton. Some, however, suggest that the Cotton Famine was mostly due to overproduction and price inflation caused by an expectation of future shortage.

- Prior to the Civil War, Lancashire companies issued surveys to find new cotton-growing countries if the Civil War were to occur and reduce American exports. India was deemed to be the country capable of growing the necessary amounts. Indeed, it helped fill the gap during the war, making up only 31 per cent of British cotton imports in 1861, but 90 per cent in 1862 and 67 per cent in 1864.
- Additionally, the main purchasers of cotton, Britain and France, began to turn to Egyptian cotton. The Egyptian government of Viceroy Isma'iltook out substantial loans from European bankers and stock exchanges. After the American Civil War ended in 1865, British and French traders abandoned Egyptian cotton and returned to cheap American exports, sending Egypt into a deficit spiral that led to the country declaring bankruptcy in 1876, a key factor behind Egypt's occupation by the British Empire in 1882.
- The South continued to be a one-crop economy until the 20th century, when the New Deal and World War II encouraged diversification. Many ex-slaves as well as poor whites worked in the sharecropping system in serf-like conditions.

Modern History

Boll Weevils

- Boll weevils, insects that entered the United States from Mexico in 1892, created 100 years of problems for the U.S. cotton industry. Many consider the boll weevil almost as important as the Civil War as an agent of change in the South, forcing economic and social changes. In total, the boll weevil is estimated to have caused $22 billion in damages. In the late 1950s, the U.S. cotton industry faced economic problems, and eradication of the boll weevil was prioritized.
- The Agricultural Research Service built the Boll Weevil Research Laboratory, which came up with detection traps and pheromone lures. The program was successful, and pesticide use reduced significantly while the boll weevil was eradicated in some areas.

Africa and India

- After the Cotton Famine, the European textile industry looked to new sources of raw cotton. The African colonies of West Africa and Mozambique provided a cheap supply. Taxes and extra-market means again discouraged local textile production. Working conditions were brutal, especially in the Congo, Angola, and Mozambique. Several revolts occurred, and a cotton black market created a local textile industry. In recent history, United States agricultural subsidies have depressed world prices, making it difficult for African farmers to compete.

- India's cotton industry struggled in the late 19th century because of unmechanized production and American dominance of raw cotton export. India, ceasing to be a major exporter of cotton goods, became the largest importer of British cotton textiles. Mohandas Gandhi believed that cotton was closely tied to Indian self-determination. In the 1920s he launched the Khadi Movement, a massive boycott of British cotton goods. He urged Indians to use simple homespun cotton textiles, khadi. Cotton became an important symbol in Indian independence.
- During World War II, shortages created a high demand for khadi, and 16 million yards of cloth were produced in nine months. The British Raj declared khadi subversive; damaging to the British imperial rule. Confiscation, burning of stocks, and jailing of workers resulted, which intensified resistance. In the second half of the 20th century, a downturn in the European cotton industry led to a resurgence of the Indian cotton industry. India began to mechanize and was able to compete in the world market.

Cotton Plant

- Cotton comes from the puffy, cotton fluff of seed pods that grow on plants that are a members of the mallow family. Cotton is ideal fore making thread and cloth because the flattened, twisted fibers that come from the seed pods interlock and bond when spun into string threads. About two months after planting, when the plant is about a foot high, flowers develop from buds called "squares."
- The flowers starts out white and then turn red on the second day after they appear. On the third day the petals fall off leaving behind undeveloped, flattened green pods called boll. The boll takes about 45 to 60 days to mature into an egg-shaped pod about an inch in diameters and an inch and a half long. Inside of each mature pod are three to five cotton-filled compartments call "locks." Each locks contains seven to ten seeds to which thousands of fluffy white cotton fibers are attached.

Cotton Cultivation

- Cotton needs lots of sun, a long hot summer and a fair amount of water to grow. It does well in dry climates and rich soils but does not do well in a climates that are too cold or too wet and requires 20 tons of water per day per acre (about the same amount of water as wheat) and needs a growing season of 200 frost-free days. Cotton depletes nutrients in the soil quickly and is often rotated with other crops. Because, it needs a lot of sun and water it is often grown in hot, dry places with water supplied by irrigation.
- The average yield is around 600 pounds of cotton per acre. Cotton-growing methods vary from region to region. In many developing countries the

work is still done mainly by hand. The fields are plowed with horse or mules or some other animal and seeds are usually planted in rows at a depth of about one or two inches deep. In the developed world, cotton is grown in vast fields and is mechanically planted, cultivated and harvested. Fields are sprayed with fertilizers and pesticides with helicopters and planes.

- They are usually planted close together as a means of support. Only when they reach a height of six centimeters are growers confident that the plant won't fail. After the plants are stronger and reach a height of ten centimeters they "chopped" to clear away weeds and excess plants. Some field have to "chopped" seven or times before the harvest.
- With cotton there is a strong risk of loss due to pest such as the boll weevil. Cotton has traditionally been grown with lots off fertilizers, pesticides and even defoliates Many of the problems that developed from DDT in the 1950s and 60s resulted from spraying huge amounts of it by plane on cotton crops. Cotton is still picked by hand in many parts of the world. One reason for this is so the plants can be harvested several times as the bolls mature at different rates. Bolls on a single plant often ripen at different times. To speed up the ripening process sometimes farmers use chemical that defoliate the plant.
- At harvest time the cotton plants are usually three to four feet high. Pickers wander through the fields with bags and pick the cotton either by "hand snapping" the entire boll, yielding cotton along with the husk and other stuff, or plucking the cotton from the boll. Skilled pickers can pick between 200 and 300 pounds of cotton a day. It is backbreaking work often done for little money. At the end of the day the hands of pickers are covered in cuts and scratches. Mechanical harvesting is done by two machines: a picker which pulls the bolls from the plant and stripper that slices the bolls and many of the leaves from the stalk.

Pests

- Cotton harvest Cotton is vulnerable to attacks by pests like boll weevils, a beetle with along nose-like appendage indigenous to the Gulf Coast area of Veracruz, Mexico where it has traditionally eaten a plant belonging to the cotton family called *Hampra.* These insects caused a great amount of destruction in the United States, arriving in Texas in 1894, spreading across much the American South by 1920 and devastated that area until it the 1950s when it began to be successfully brought under control with insecticides such as DDT.
- An early report of the insects destructive powers sent by a farmer from Texas to Washington read: "The 'Top' crop of cotton of this section has been very much damaged and in some cases almost entirely destroyed by a peculiar weevil or bug which by some means destroys the squares

of the small bolls. Our farmer can combat the cotton worm but are at loss to know what to do to overcome this pest."

- Creating a cotton module With their long snout boll weevils pierce the flower buds or "squares" and eat the pollen inside. They also puncture the immature fruit of "bolls" and consume the developing cotton inside. But most destructively, the female drills deep inside the bolls and squares and deposits eggs. These hatch, releasing larvae that eat the plant's reproductive structure, causing more damage to the squares and bolls. By the time they are finished a cotton field can be as much as 80 per cent destroyed.
- Boll weevils still cause around $150 million worth of damage a year in the United States but have largely been brought under control there thanks to traps bated with boll weevil pheromes and Malathion, a strong, effective chemical insecticide that has to be used sparingly because of the environmental damage it can cause. The boll weevil remains a threat in Mexico and Central America and is making advances in South America. The cotton leafworm is another cotton pest that comes from the Veracruz area of Mexico.

Production

- Cotton is regarded as an industrial crop. Modern mills are clean and have quiet machines. Some "virtually run themselves." Older factories found in places like India rely on machinery brought from England 100 years ago and are noisy, dirty, dangerous places filled inhalable dust and lint and mill laborers performing mind-numbing and body-destroying chores.
- The first thing that is done after the cotton is harvested is the separation of the seeds from the cotton with a cotton gin consisting of hooked-tooth saws that tear the fibers from the seeds. Modern gins also remove moisture, impurities and dust from the fibers. Afterwards, the free fibers, called lint, are sucked away into a collection area. The seeds are saved to use for next year's crop or crushed to make oil for margarine, cooking oil or cosmetics. The cottons fibers are pressed into bales weighing around 500 pounds.
- At a cotton mills picking machines clean the cotton and make it into rolls called laps. Carding machines, cylinders with wire points, straightens the fibers and mold them into finger-thick, ropelike strands called slivers. Drawing frames draw out several slivers at one time and combine them into single strands. Riving machines then twists and further draw out the cotton into a thinner strand. Spinning machine then draw and twist the strands into yarns of desired sizes and widens them on bobbins for weaving.

- ☆ Yarns of cotton are woven together on looms with warps (lengthwise yarns) on rollers. Every other warp thread is lifted all at once while a shuttle drives wefts (crosswise yarn) under the even warp threads and over the odd ones. Rollers shift so that the weft carried by the shuttle on the return journey goes over the even warp threads and under the odd ones. The shuttle can make 200 or more trips a minute. Afterwards the unfinished "gray goods" fabric is taken to a converter for final processing. This includes bleaching, dying, preshrinking and printing the fabric. The fabric can be combed into denim and given a finishing of crepe, glazed chintze or water resistance.

Chapter 12

Gardening in Ancient and Medieval Period: Arbori-horticulture Orchard

Indian History of Gardening

> *Horticulture is the study of plants. It involves plant cultivation for food, utility, beauty and recreation. A horticulturist is an individual who studies, works with or makes enhancements to the cultivation of flowers and plants for ornamental use. Horticulturists take care of greenery, gardens, golf courses, parks and all other forms of landscaping.*

- India has a long history of flowering plants. In the era of Mahabharata, there was famous tree named Kadamba, which associated with Lord Krishna. Vatsyana (A.D. 300-400) described four kinds of gardens, which were made for the queens, kings, courtiers and ministers. Famous poet Bana Bhatta described the number of flowering plants in his famous book the Harsh Charita. These flowers plants were growing in the gardens. At that time, water pools built with red lotus and blue water lilies. Status of gardening had mentioned in Ramayana written by Valmiki and Tulsidas. At that time, Ayodhya city was having wide streets, large houses, noble palaces, richly decorated temples and gardens. These gardens were planted with fruit trees, flowers; lakes were full of lotuses and different kinds of birds.
- In the period of Lord Buddha, the life of Buddha was associated with a number of trees from birth to his nirvana. He was born in 563 B.C. under the tree of Ashoka at Lumbini. Birth place of Buddha has been described by Hiuen Tsang who visited the place in 630 A.D., there was bathing tank

of Sakya Muni filled with clear water, lotuses and lilies when Lord Buddha visited Vaishali, Amrapali presented a park known as Amrvana which was dominated by flowering trees. Buddhist was planting trees and flowering plants on a large scale for making surrounding peaceful, a place ideal for meditation.

☆ When Aryans came in India about 1600 B.C., at that time, the country was called as Aryavrta, which means the country of lotus and sunshine because the lakes were studded with lotus flowers and there were wide-open spaces. Therefore, the lotus being a native of India found everywhere. Aryans started the use of flowers in religious and social ceremonies. They appreciated the beauty of flowering plants, lakes, mountains, flowers like Kamal, Champa, Madhavi, Bela, Chameli, Rukmani, *etc.*

> *Arboriculture is the study of trees. An arborist is considered to be an individual who studies trees and the proper ways to take care of them. They are also sometimes called tree surgeons. Arborist's are needed to evaluate the condition of trees, make recommendations for their care, and provide services to keep them healthy and thriving for years to come.*

☆ The history of systematic gardening in India is as old as civilization of Indus of Harappa, which existed between 2400 B.C. and 1750 B.C. At that time, people were living in well-planned roads cut across one another almost at right angles. There are many evidences found that trees and ornamental plants were associated with the Harappa civilization. In Mogul era, Babar had founder of gardens. He made gardens at Panipat and Agra. Mogul gardens are synonymous of formal style of gardening. Grand Trunk Road from Lahore to Calcutta made by Sher Shah Suri and planted shady trees along both sides of the roads. Akbar has made Fatehpur Sikri (Agra) garden. There are so many gardens build by Mogul emperors.

☆ During the British period, there was a lot of activity in gardening by Britishers and Indian kings. King Hyder Ali established most famous Lal Bag garden at Bangalore. In North India, Maharaja Ranjit Singh made garden at Amritsar. Britishers had managed well gardens in India. They imports plants from England. Britishers established Royal-Agri-Horticulture Societies and Botanical Gardens in India *i.e.* Royal Agri-Horticultural Society Garden, Calcutta., Lloyd botanical garden, Darjeeling, Botanical garden, Saharanpur, National Botanical Garden, Lucknow, Botanical garden of the forest research garden, Ootacmund; *etc.*

☆ There have been changes in the field of gardening during post independence period. An effort has made for public gardens in big cities for improving environment. Several gardens in different cities have been providing recreational facilities. Some important gardens are Buddha Jayanti garden Delhi, Rose garden, Chandigarh *etc.* The gardens has made along with mega highway all over India recently. Because of these stages of gardening, have increased flowering habits among Indians. After globalization, modern trends are transferring from one nation to another, though the new trends arrive in India. Therefore, last two decades there has been raising commercial floriculture in India.

World History of Gardening

- The evidence about the existence of gardens in ancient Egypt is archaeological. Different portions of the plants recovered from tombs and a number of plants painted on the tombs. These painting show that the Egyptians had religious. There was Temple Garden at Karnak dating from the region of Tuth Mosis II, about 1500 B.C. Queen Hatsheput (1505-1483 B.C.) built a Temple Garden on a hill. It was a terrace garden with the temple on the uppermost terrace. It also indicate that the idea and technique of garden making from Mesopotamia to Egypt.
- China is one of the centres of humans evolution. The great poet Li-Po (A.D.705-762) of Tang period has noted importance of garden and nature in his poem and we seen by this evidence China has the origin of the flora.
- The Japanese are great lovers of plants and flowers. The first garden was constructed in Kyoto for holding garden parties dating A.D. 794 in the period of Heian.
- Iran apart from love, war and hunting they were founder of gardening from 558- 323 B.C. Cyrusi who was the first Persian Gardner in that period. He planted a garden of Sandins in Lydia, had set the design of the Chahav Bagh.
- The first park in Greece was dedicated to Diana in the period 434-356 B.C. In this park, fruit trees were planted around the temple. The king Xenophon got his inspiration from Persian garden in that period.
- Italy, one of the heartland of the Rome Empire. Italy was the state of agriculture in third and second centuries B.C. At that time, few farmers were horticulturist and growers of gardens. In Arab countries, during A.D. 1058 to A.D.1259 that a galaxy of men of genius flourished that is Al Bakri and Idrisi.
- In modern age Europe made several gardens, *i.e.* Dahlias and Zinnias are the gifts of Mexican gardeners to the world. Thus, it would be seen that the countries of new world has sophisticated gardens.

Mauryan Period

- The Mauryas in the 4th to 3rd century BC, there has been vast secular literature and texts, both Vedic and post-Vedic, like Vedas, Brahamanas, Aranyakas, Upanishadas, Sutras, Smritis, Mahakavyas, Puranas, Buddhists texts (Jataka) and Jain literature (Sutras).

- The sages of the Upanishadas have mentioned the Cosmic Tree rooted in the Brahman, the ultimate, whose branches are space, wind, fire, water and earth.
- Kalpavrksa and kalpalata are mythological tree and creeper, not described in the Vedic literature. which have been a part of folk cult in Hindu mythology. Kalpavrksa is mentioned in Ramayana, Mahabharata, Jatakas, Divyavadana and the Jain Sutras. In Brahamanical religion, vata (*Ficus benghalensis*) was identified with Shiva, asvattha (*Ficus religiosa*) with Vishnu, lotus with Surya (Sun) and nine leaves of nine trees (navapatrika).
- Descriptions of trees in the Rigveda (3700-2000 BC), the Ramayana (1200-1000 BC), The Mahabharata as well as other literature by Shudraka (100 BC), Kalidasa (c. 57 BC), Ashvaghosha (100 AD), Vatsyayana (300-400 AD) and Sarangdhara (1300 AD). The gardening and kinds of gardens were described by Sarangdhara and Vatsyayana.
- In the Ramayana, mention is made of Panchavati, in which Sita was held captive. Ashoka trees (*Saraca asoca*) were predominant in this garden. In the Panchavati garden, five trees were planted which are:
  - Asvattha (*Ficus religiosa*) on the east side,
  - Bilva (*Aegle marmelos*) on the north,
  - Banyan (*Ficus benghalensis*) on the west,
  - Amla (*Emblica officinalis*) on the south, and
  - Ashoka (*Saraca asoca*) on the southeast.
- The layout of gardens and parks and artificial lakes in the city of Indraprastha is given in the Sabha-Parva of the Mahabharata. Several trees, such as *Saraca asoca, Terminalia arjuna, Mesua ferrea, Ficus benghalensis, F. religosa, Michelia champaka, Butea monosperma* and *Cassia fistula*, have been also describe in the Ramayana.
- Lord Buddha was born under the pipal tree in a garden. The bodhi tree, under which the Buddha attained nirvana, is sacred to Buddhists.
- The trees described in Buddhist texts are:
  - Asvattha (*Ficus religiosa*),
  - Banyan (*Ficus benghalensis*),
  - Udumbara (*Ficus glomerata*),
  - Patali (*Bignonia suaveolens*),
  - Sala (*Shorea robusta*) and
  - Sirisa (*Acacia sarisa*).

Vedic Times

The plants described in the Vedic times were:

- Tulsi (*Ocimum sanctum*)

- ☆ Sami (*Prosopis ceneraria*)
- ☆ Banyan (*Ficus benghalensis*)
- ☆ Soma (*Sarcostemma acidum*)
- ☆ Udumbara (*Ficus glome rata*)
- ☆ Bilva (*Agele marmelos*)
- ☆ Khadira (*Acacia catechu*)
- ☆ Neem (*Azadirachta indica*)
- ☆ Palasa (*Butea monosperma*)
- ☆ Lotus (*Nelumbo nucifera*)
- ☆ Pipal (*Ficus religiosa*)

Other trees and plants of the Vedic and post-Vedic period

- ☆ Coconut (*Cocos nucifera*)
- ☆ Rudraksha (*Elacocarpus sphaericus*)
- ☆ Shal (*Shorea robusta*)
- ☆ Asoka (*Saraca asoca*)
- ☆ Snuhi (*Euphorbia neriifolia*)
- ☆ Madhavi lata (*Hiptage madablota*)
- ☆ Amalka (*Emblica officinalis*)
- ☆ Cotton tree (*Bombax ceiba*)
- ☆ Mango, amra (*Mangifera indica*)
- ☆ Banana (*Musa paradisiaca*)
- ☆ Ber (*Zizyphus mauritina*)
- ☆ Kadamba (*Anthocephalus cadamba*)
- ☆ Bahira (*Terminalia belli rica*)
- ☆ Arjuna (*Terminalia arjuna*)
- ☆ Tili (*Sesamum indicum*)
- ☆ Imli (*Tamarindus indica*)
- ☆ Parijata (*Nycanthes arbortristis*)
- ☆ Tinduku (*Diospyros peregrina*)
- ☆ Nalaka (*Arundo donax*)
- ☆ Jivaka (*Putranjiva roxburghii*)
- ☆ Mandara (*Erythrina variegata*)
- ☆ Paan (*Piper betle*)
- ☆ Ketaki (*Pandanus odoratissimus*)
- ☆ Amarphal (*Monstera deliciosa*)
- ☆ Gaduchi (*Cocculus cordifolius*)

Tree Worship

- ✰ The tree motifs have been found in the art of Indus Valley, Mauryan ring stones, and gateways and railings of stupas at Bharhut, Bodhgaya, Sanchi, Amaravati and Nagajunakonda, Mauryan relief sculptures.
- ✰ Ancient sculptures and architecture of Mathura (Kanishka period, AD 78-101) and Ajanta frescoes (AD 100-600) also bear testimony to the importance of plants and flowers. The relationship of trees with the Brahamanical and Buddhist gods and goddesses and Jain Tirthankaras in Indian art date back.
- ✰ Trees and flowers have been also delineated in ancient coins found at the pre-Mauryan site, Sugh, Taxila, Ayodhya during Mitra Kings, Kausambi and Mathura and also of the Andhra dynasty and Pandyan territory. The ancient Sanskrit and other literature and texts, mythological epics and legends, paintings, cave murals and frescoes, sculptures, architecture, folklores and tribal arts and crafts provide evidences of the kind of plants and trees.
- ✰ The Vrikshayurveda and arbori-horticulture, and usefulness of forests and gardens were well-known in ancient India. The utilitarian qualities of trees and plants for food, medicine, shelter, shadow and fuel, and the relationship of trees with fertility were also known to ancient Indians and they were concerned with the conservation of trees and biodiversity in nature and ecological balance.
- ✰ The concept of identifying trees with gods and goddesses, and threats and punishments against the destruction of useful trees helped to save the trees and flora.

Mughal Period

- ✰ Gardening in India by the Mughal rulers beginning with Babur, many plant species were brought by them from Persia and Central Asia where herbaceous and bulbous flowers were already under cultivation. Many of these have been described in autobiographies and other books written during those days. Besides, in Mughal paintings also we find illustrations of many flowers. These have also been used to illustrate the borders of those paintings. In the book Bagh-I- Wafa.
- ✰ In the 16th and 17th centuries AD, Mughal gardens were developed in Agra, Delhi, Pinjore, Srinagar, Kashmir and a few other Indian places.
- ✰ The most important Mughal gardens are:
  - ❀ Brindavan Gardens, Mysore, Karnataka
  - ❀ Hamayun's tomb, Delhi
  - ❀ Khusro Bagh, Allahabad, Uttar Pradesh
  - ❀ The Presidential (Rashtrapati Bhawan) Garden, New Delhi

- Mughal Gardens of Kashmir, Jammu and Kashmir
- Shalimar and Nishat Gardens, Srinagar
- Taj Mahal Garden, Agra, Uttar Pradesh
- The Mughal Garden, Pinjore, Haryana

☆ The rose was introduced into our country via the port of Bussorah by Babur 1526. Jehangir and Nurjehan were ardent lovers of the rose and encouraged rose growing in gardens.

☆ The most important plants introduced in Kashmir from Persia by the Mughal ruler, Jehangir in 1619 when he laid out the famous Shalimar Bagh in Srinagar, were the majestic Chinar tree (*Platanus orientalis*), the cypress (*Cupresus sempervirens*) and the weeping willow (*Salix babylonica*), and flowers like rose, narcissus, daffodil *etc.*

European Period

☆ Later, mainly Englishmen and the Portuguese introduced many species. Missionaries and priests, civil servants and individual amateur, gardeners mostly brought these in. One of the important missionaries who introduced a number of exotic plants was Dr Firminger, an Englishman, who wrote a book on gardening, giving descriptions of various species of flowers in 1863. The book titled 'Firminger's Manual of Gardening in India' is an authoritative reference book on ornamental flowering plants even today.

☆ With the establishment of Government Botanic Gardens by the British rulers during 18th and 19th centuries, such as:

- Lalbagh Botanical Garden, Bangalore (1760);
- The Government Botanic Garden, Saharan pur (1779);
- The Indian Botanic Garden, Sibpur, Calcutta (1787);
- The Lloyd Botanic Garden, Darjeeling (1878);
- The Government Botanic Garden, Oatacamund (1884).

☆ Some important and rare flowers of India are:

- *Agapetes auriculata*
- *Corydalis govaniana*
- *Dendrobium chrysanthum*
- *Dendrobium nobile*
- *Geranium wallichianum*
- *Katherinea ampla*
- *Meconopsis aculeate*
- *Notholirion thomsonianum*
- *Nepenthes khasiana*

- *Rhododendron macabeanum*
- *Rhododendron hodgsonii*
- *Rhododendron thomsonii.*

History of Ornamentals Plants

- Flowers have always remained an integral part of the social fabric of human life. In ancient India, flowers were extensively used for offerings to Gods and deities, for making garlands, landscaping of gardens, in various rituals, ceremonies, feasts and festivals.
- In Vedas, many flower-bearing plants were mentioned such as Kamal (Lotus), Kumud (White Waterlily), Champa (*Magnolia champaca*), Chameli (Jasmine), Madhavi (*Hiptage benghalensis*), Neel Kamal (Blue Waterlily), Palasa (*Butea monosperma*) *etc.*
- Flowers have been cultivated in India from time immemorial. Ornamental plants have been an integral part of Indian Civilizations with their varied uses in religious and socio-cultural rituals. Many wild ornamental plants were domesticated and cultivated in gardens and backyards for their varied utilizations.

Table 12.1: Some Wild Native Plants of India with high Ornamental Values

Sl.No.	*Taxa*	*Family*
1.	*Abroma augusta* (L.) L.f.	Sterculiaceae
2.	*Aerides odorata* Lour.	Orchidaceae
3.	*Agapetes odontocera* (Wight) Benth. and Hook.f.	Ericaceae
4.	*Anemone tetrasepala* Royle	Ranunculaceae
5.	*Arisaema tortuosum* (Wall.) Schott	Araceae
6.	*Barleria cristata* L.	Acanthaceae
7.	*Bauhinia glauca* (Benth.) Benth.	Fabaceae
8.	*Bauhinia variegata* L.	Fabaceae
9.	*Begonia picta* Sm.	Begoniaceae
10.	*Burmannia disticha* L.	Burmanniaceae
11.	*Carduus edelbergii* Rech.f.	Asteraceae
12.	*Clerodendrum chinense* (Osbeck) Mabb.	Lamiaceae
13.	*Coelogyne cristata* Lindl.	Orchidaceae
14.	*Crinum latifolium* L.	Amaryllidaceae
15.	*Cymbidium bicolor* Lindl.	Orchidaceae
16.	*Delphinium malabaricum* (Huth) Munz	Ranunculaceae
17.	*Euphorbia sikkimensis* Boiss.	Euphorbiaceae
18.	*Geranium himalayense* Klotzsch	Geraniaceae
19.	*Gerbera gossypina* (Royle) Beauverd	Asteraceae

Sl.No.	Taxa	Family
20.	*Globba sessiliflora* Sims	Zingiberaceae
21.	*Hedychium coccineum* Buch.-Ham. ex Sm.	Zingiberaceae
22.	*Hoya carnosa* (L.f.) R.Br.	Apocynaceae
23.	*Hypericum choisianum* Wall. ex N.Robson	Clusiaceae
24.	*Impatiens monticola* Hook.f.	Balsaminaceae
25.	*Incarvillea emodi* (Royle ex Lindl.) Chatterjee	Bignoniaceae
26.	*Iris kemaonensis* Wall. ex D.Don	Iridaceae
27.	*Ixora coccinea* L.	Rubiaceae
28.	*Lobelia clavata* E.Wimm.	Campanulaceae
29.	*Luculia pinceana* Hook.	Rubiaceae
30.	*Magnolia hodgsonii* (Hook.f. and Thomson) H.Keng	Magnoliaceae
31.	*Melastoma malabathricum* L.	Melastomataceae
32.	*Mussaenda macrophylla* L.	Rubiaceae
33.	*Osbeckia stellata* Buch.-Ham. ex Ker Gawl.	Melastomataceae
34.	*Passiflora foetida* L.	Passifloraceae
35.	*Pavetta crassicaulis* Bremek.	Rubiaceae
36.	*Pholidota articulata* Lindl.	Orchidaceae
37.	*Prunus cerasoides* Buch.-Ham. ex D.Don	Rosaceae
38.	*Rhododendron ciliatum* Hook. f.	Ericaceae
39.	*Saurauia napaulensis* DC.	Actinidiaceae
40.	*Schima wallichii* Choisy	Theaceae
41.	*Tacca integrifolia* Ker Gawl.	Dioscoreaceae
42.	*Tamilnadia uliginosa* (Retz.) Tirveng and Sastre	Rubiaceae
43.	*Thunbergia mysorensis* (Wight) T. Anderson	Acanthaceae
44.	*Thunia alba* (Lindl.) Rchb.f.	Orchidaceae
45.	*Woodfordia fruticosa* (L.) Kurz	Lythraceae

- ✰ During the Mughal period, there was a tremendous change with wide scale use of many ornamental plants in the landscaping of gardens across the country. Many monumental gardens were established and several native flowering plants were used to beautify them.
- ✰ With the arrival of European Colonizers in 17th Century and the subsequent colonial rule thereafter, there was influx of exotic species in gardens and landscapes of the country. Beautiful exotic species were extensively used in landscaping of gardens, parks and avenues.
- ✰ Flowering plants from across the globe were introduced and cultivated to support the booming floriculture market. In recent times, India has become one of the leading players in global markets.

Chapter 13

Vegetable Farming: Floriculture, Perfumes and Medicinal Plants

An Overview

- ☆ The word *vegetable* was first recorded in English in the early 15th century. It comes from Old French, and was originally applied to all plants; the word is still used in this sense in biological contexts. It derives from Medieval Latin *vegetabilis* "growing, flourishing" (*i.e.* of a plant), a semantic change from a Late Latin meaning "to be enlivening, quickening". The meaning of "vegetable" as a "plant grown for food" was not established until the 18th century. In 1767, the word was specifically used to mean a "plant cultivated for food, an edible herb or root". The year 1955 saw the first use of the shortened, slang term "veggie". As an adjective, the word *vegetable* is used in scientific and technical contexts with a different and much broader meaning, namely of "related to plants" in general, edible or not–as in *vegetable matter, vegetable kingdom, vegetable origin, etc.*
- ☆ All societies and ethnic groups eat vegetables because they are essential for maintaining human health. In simple terms, modern vegetable science deals with growing herbaceous plants for human consumption to meet basic nutritional needs. As the world's population grows, the demand for vegetables will continue to grow as well. Vegetable science, sometimes called olericulture, is one of the most dynamic and important fields of the agricultural sciences. The importance of vegetables has never been greater.

- ☆ Vegetables are a horticultural food crop. Other horticultural food crops include small fruits and tree fruits, which are usually grown as perennials. Vegetable crops may be either annuals or perennials. From a production standpoint, a vegetable crop may be defined as a high-value crop that is intensively managed and requires special care after harvest "vegetable" is a term based on the usage of herbaceous plants or portions of plants that are eaten whole or in part, raw or cooked, generally with an entree or in a salad but not as a dessert, that are intensively managed and may require special care after harvest to maintain quality.

Vegetable Farming

History

- ☆ Before the advent of agriculture, humans were hunter-gatherers. They foraged for edible fruit, nuts, stems, leaves, corms, and tubers, scavenged for dead animals and hunted living ones for food. Forest gardening in a tropical jungle clearing is thought to be the first example of agriculture; useful plant species were identified and encouraged to grow while undesirable species were removed. Plant breeding through the selection of strains with desirable traits such as large fruit and vigorous growth soon followed. While the first evidence for the domestication of grasses such as wheat and barley has been found in the Fertile Crescent in the Middle East, it is likely that various peoples around the world started growing crops in the period 10,000 BC to 7,000 BC.
- ☆ Subsistence agriculture continues to this day, with many rural farmers in Africa, Asia, South America, and elsewhere using their plots of land to produce enough food for their families, while any surplus produce is used for exchange for other goods.
- ☆ Throughout recorded history, the rich have been able to afford a varied diet including meat, vegetables and fruit, but for poor people, meat was a luxury and the food they ate was very dull, typically comprising mainly some staple product made from rice, rye, barley, wheat, millet or maize.
- ☆ The addition of vegetable matter provided some variety to the diet. The staple diet of the Aztecs in Central America was maize and they cultivated tomatoes, avocados, beans, peppers, pumpkins, squashes, peanuts, and amaranth seeds to supplement their tortillas and porridge. In Peru, the Incas subsisted on maize in the lowlands and potatoes at higher altitudes. They also used seeds from quinoa, supplementing their diet with peppers, tomatoes, and avocados.
- ☆ In Ancient China, rice was the staple crop in the south and wheat in the north, the latter made into dumplings, noodles, and pancakes. Vegetables used to accompany these included yams, soybeans, broad beans, turnips, spring onions, and garlic. The diet of the ancient Egyptians was based on

bread, often contaminated with sand which wore away their teeth. Meat was a luxury but fish was more plentiful. These were accompanied by a range of vegetables including marrows, broad beans, lentils, onions, leeks, garlic, radishes, and lettuces.

☆ The mainstay of the Ancient Greek diet was bread, and this was accompanied by goat's cheese, olives, figs, fish, and occasionally meat. The vegetables grown included onions, garlic, cabbages, melons, and lentils. In Ancient Rome, a thick porridge was made of emmer wheat or beans, accompanied by green vegetables but little meat, and fish was not esteemed. The Romans grew broad beans, peas, onions and turnips and ate the leaves of beets rather than their roots.

Table 13.1: Some Common Vegetables

Vegetable Image	*Scientific Name*	*Parts Used*	*Origin*	*Cultivars*
	Brassica oleracea	Leaves, axillary buds, stems, flower heads	Europe	Cabbage, Brussels sprouts, cauliflower, broccoli, kale, kohlrabi, red cabbage, Savoy cabbage, Chinese broccoli, collard greens
	Brassica rapa	Root, leaves	Asia	Turnip, Chinese cabbage, napa cabbage, bok choy
	Raphanus sativus	Root, leaves, seed pods, seed oil, sprouting	South-eastern Asia	Radish, daikon, seedpod varieties
	Daucus carota	Root, leaves, stems	Persia	Carrot
	Pastinaca sativa	Root	Eurasia	Parsnip

Vegetable Image	*Scientific Name*	*Parts Used*	*Origin*	*Cultivars*
	Beta vulgaris	Root, leaves	Europe and Near East	Beetroot, sea beet, Swiss chard, sugar beet
	Lactuca sativa	Leaves, stems, seed oil	Egypt	Lettuce, celtuce
	Phaseolus vulgaris *Phaseolus coccineus* *Phaseolus lunatus*	Pods, seeds	Central and South America	Green bean, French bean, runner bean, haricot bean, Lima bean
	Vicia faba	Pods, seeds	Mediterranean and Middle East	Broad bean
	Pisum sativum	Pods, seeds, sprouts	Mediterranean and Middle East	Pea, snap pea, snow pea, split pea
	Solanum tuberosum	Tubers	South America	Potato
	Solanum melongena	Fruits	South and East Asia	Eggplant (aubergine)

Vegetable Image	Scientific Name	Parts Used	Origin	Cultivars
	*Solanum lycopersicum*	Fruits	South America	Tomato, see list of tomato cultivars
	*Cucumis sativus*	Fruits	Southern Asia	Cucumber, see list of cucumber varieties
	*Cucurbita* spp.	Fruits, flowers	Mesoamerica	Pumpkin, squash, marrow, zucchini (courgette), gourd
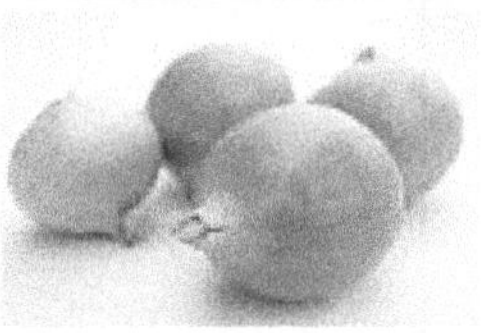	*Allium cepa*	Bulbs, leaves	Asia	Onion, spring onion, scallion, shallot, see list of onion cultivars
	*Allium sativum*	Bulbs	Asia	Garlic
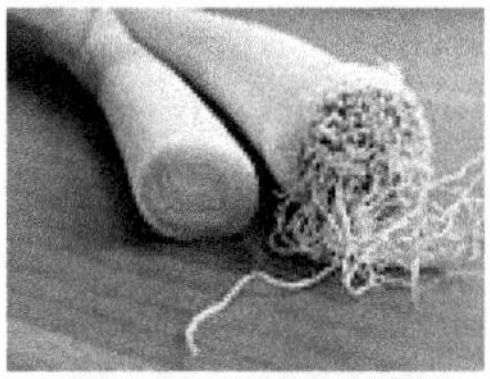	*Allium ampeloprasum*	Leaf sheaths	Europe and Middle East	Leek, elephant garlic
	*Capsicum annuum*	Fruits	North and South America	Pepper, bell pepper, sweet pepper

Vegetable Image	*Scientific Name*	*Parts Used*	*Origin*	*Cultivars*
	*Spinacia oleracea*	Leaves	Central and southwestern Asia	Spinach
	*Dioscorea* spp.	Tubers	Tropical Africa	Yam
	*Ipomoea batatas*	Tubers, leaves, shoots	Central and South America	Sweet potato, see list of sweet potato cultivars
	*Manihot esculenta*	Tubers	South America	Cassava

Important Role in Human Nutrition

☆ Vegetables play an important role in human nutrition. Most are low in fat and calories but are bulky and filling. They supply dietary fiber and are important sources of essential vitamins, minerals, and trace elements. Particularly important are the antioxidant vitamins A, C, and E. When vegetables are included in the diet, there is found to be a reduction in the incidence of cancer, stroke, cardiovascular disease, and other chronic ailments. Research has shown that, compared with individuals who eat less than three servings of fruits and vegetables each day, those that eat more than five servings have an approximately twenty per cent lower risk of developing coronary heart disease or stroke. The nutritional content of vegetables varies considerably; some contain useful amounts of protein though generally they contain little fat, and varying proportions of vitamins such as vitamin A, vitamin K, and vitamin B_6; provitamins; dietary minerals; and carbohydrates.

- However, vegetables often also contain toxins and antinutrients which interfere with the absorption of nutrients. These include α-solanine, α-chaconine, enzyme inhibitors (of cholinesterase, protease, amylase, *etc.*), cyanide and cyanide precursors, oxalic acid, tannins and others. These toxins are natural defenses, used to ward off the insects, predators and fungi that might attack the plant. Some beans contain phytohaemagglutinin, and cassava roots contain cyanogenic glycoside as do bamboo shoots. These toxins can be deactivated by adequate cooking. Green potatoes contain glycoalkaloids and should be avoided.
- Fruit and vegetables, particularly leafy vegetables, have been implicated in nearly half the gastrointestinal infections caused by norovirus in the United States. These foods are commonly eaten raw and may become contaminated during their preparation by an infected food handler. Hygiene is important when handling foods to be eaten raw, and such products need to be properly cleaned, handled, and stored to limit contamination.

Cultivation

- Vegetables have been part of the human diet from time immemorial. Some are staple foods but most are accessory foodstuffs, adding variety to meals with their unique flavors and at the same time, adding nutrients necessary for health. Some vegetables are perennials but most are annuals and biennials, usually harvested within a year of sowing or planting. Whatever system is used for growing crops, cultivation follows a similar pattern; preparation of the soil by loosening it, removing or burying weeds, and adding organic manures or fertilisers; sowing seeds or planting young plants; tending the crop while it grows to reduce weed competition, control pests, and provide sufficient water; harvesting the crop when it is ready; sorting, storing, and marketing the crop or eating it fresh from the ground.
- Different soil types suit different crops, but in general in temperate climates, sandy soils dry out fast but warm up quickly in the spring and are suitable for early crops, while heavy clays retain moisture better and are more suitable for late season crops. The growing season can be lengthened by the use of fleece, cloches, plastic mulch, polytunnels, and greenhouses. [31] In hotter regions, the production of vegetables is constrained by the climate, especially the pattern of rainfall, while in temperate zones, it is constrained by the temperature and day length.
- On a domestic scale, the spade, fork, and hoe are the tools of choice while on commercial farms a range of mechanical equipment is available. Besides tractors, these include ploughs, harrows, drills, transplanters, cultivators, irrigation equipment, and harvesters. New techniques are changing the cultivation procedures involved in growing vegetables with

computer monitoring systems, GPS locators, and self-steer programs for driverless machines giving economic benefits.

Harvesting

- When a vegetable is harvested, it is cut off from its source of water and nourishment. It continues to transpire and loses moisture as it does so, a process most noticeable in the wilting of green leafy crops. Harvesting root vegetables when they are fully mature improves their storage life, but alternatively, these root crops can be left in the ground and harvested over an extended period.
- The harvesting process should seek to minimise damage and bruising to the crop. Onions and garlic can be dried for a few days in the field and root crops such as potatoes benefit from a short maturation period in warm, moist surroundings, during which time wounds heal and the skin thickens up and hardens. Before marketing or storage, grading needs to be done to remove damaged goods and select produce according to its quality, size, ripeness, and color.

Postharvest Storage and Preservation

Postharvest Storage

- All vegetables benefit from proper post harvest care. A large proportion of vegetables and perishable foods are lost after harvest during the storage period. These losses may be as high as thirty to fifty per cent in developing countries where adequate cold storage facilities are not available. The main causes of loss include spoilage caused by moisture, moulds, micro-organisms, and vermin.
- Storage can be short-term or long-term. Most vegetables are perishable and short-term storage for a few days provides flexibility in marketing. During storage, leafy vegetables lose moisture, and the vitamin C in them degrades rapidly. A few products such as potatoes and onions have better keeping qualities and can be sold when higher prices may be available, and by extending the marketing season, a greater total volume of crop can be sold. If refrigerated storage is not available, the priority for most crops is to store high-quality produce, to maintain a high humidity level, and to keep the produce in the shade.
- Proper post-harvest storage aimed at extending and ensuring shelf life is best effected by efficient cold chain application. Cold storage is particularly useful for vegetables such as cauliflower, eggplant, lettuce, radish, spinach, potatoes, and tomatoes, the optimum temperature depending on the type of produce. There are temperature-controlling technologies that do not require the use of electricity such as evaporative cooling. Storage of fruit and vegetables in controlled atmospheres with

high levels of carbon dioxide or high oxygen levels can inhibit microbial growth and extend storage life.

- ✰ The irradiation of vegetables and other agricultural produce by ionizing radiation can be used to preserve it from both microbial infection and insect damage, as well as from physical deterioration. It can extend the storage life of food without noticeably changing its properties.

Preservation

- ✰ The objective of preserving vegetables is to extend their availability for consumption or marketing purposes. The aim is to harvest the food at its maximum state of palatability and nutritional value, and preserve these qualities for an extended period. The main causes of deterioration in vegetables after they are gathered are the actions of naturally-occurring enzymes and the spoilage caused by micro-organisms. Canning and freezing are the most commonly used techniques, and vegetables preserved by these methods are generally similar in nutritional value to comparable fresh products with regards to carotenoids, vitamin E, minerals. and dietary fiber.
- ✰ Canning is a process during which the enzymes in vegetables are deactivated and the micro-organisms present killed by heat. The sealed can excludes air from the foodstuff to prevent subsequent deterioration. The lowest necessary heat and the minimum processing time are used in order to prevent the mechanical breakdown of the product and to preserve the flavor as far as is possible. The can is then able to be stored at ambient temperatures for a long period.
- ✰ Freezing vegetables and maintaining their temperature at below −10°C (14°F) will prevent their spoilage for a short period, whereas a temperature of −18°C (0°F) is required for longer-term storage. The enzyme action will merely be inhibited, and blanching of suitably sized prepared vegetables before freezing mitigates this and prevents off-flavors developing. Not all micro-organisms will be killed at these temperatures and after thawing the vegetables should be used promptly because otherwise, any microbes present may proliferate.
- ✰ Traditionally, sun drying has been used for some products such as tomatoes, mushrooms, and beans, spreading the produce on racks and turning the crop at intervals. This method suffers from several disadvantages including lack of control over drying rates, spoilage when drying is slow, contamination by dirt, wetting by rain, and attack by rodents, birds, and insects. These disadvantages can be alleviated by using solar powered driers. The dried produce must be prevented from reabsorbing moisture during storage.

☆ High levels of both sugar and salt can preserve food by preventing micro-organisms from growing. Green beans can be salted by layering the pods with salt, but this method of preservation is unsuited to most vegetables. Marrows, beetroot, carrot, and some other vegetables can be boiled with sugar to create jams. Vinegar is widely used in food preservation; a sufficient concentration of acetic acid prevents the development of destructive micro-organisms, a fact made use of in the preparation of pickles, chutneys and relishes. Fermentation is another method of preserving vegetables for later use. Sauerkraut is made from chopped cabbage and relies on lactic acid bacteria which produce compounds that are inhibitory to the growth of other micro-organisms.

Floriculture

☆ Floriculture is the growing of cut flowers, potted flowering and foliage plants, and bedding plants in greenhouses and/or in fields. There are several thousand different species of flowers and plants that are grown as commercial crops. Cut flowers include such crops as roses, freesia, alstromeria and snapdragons. Some of the favourite flowering potted plants that are available year-round are African violets, orchids, cyclamen and potmums (potted Chrysanthemums). Some seasonal flowering plants are an important part of our traditions, for example, poinsettias for Christmas and Easter lilies for Easter.

Importance of Floriculture

Besides food and nutritional security, the aesthetic value is also equally important for our daily lively hood as well as for environmental purity. Floriculture is important from the following point of view:

1. Economic point of view
2. Aesthetic point of view
3. Social point of view

History of Flowers

☆ In the middle ages a Rose was suspended from the ceiling of a council chamber, pledging all present to secrecy, or sub Rosa, "under the Rose". Evidence of flowers dating back to the prehistoric period have been discovered through Flower Fossils. There are traces of association of flowers with humans during the paleolithic age.

☆ Archaeologists discovered skeletons of a man, two women and an infant buried together in soil containing pollen of flowers in a cave in Iraq. This association of flowers with the cave dwelling Neanderthals of the Pleistocene epoch is indicative of the role of flowers in burial rituals. Analysis of the sediment pollen concentrated in batches, implied that possible bunches of flowers had been placed on the grave. Closer examination of the flower pollen enabled scientists to identify many flowers that were present, all of which had some therapeutic properties. The use of flowers also testify abundant floral varieties available at that time.

- People have used flowers to express their feelings, enhance their surroundings, and to commemorate important rituals and observances. All forms of art, depict the use of flowers: music, books, paintings, sculpture, ceramics, tapestries, *etc*. Some of the most opulent examples of source material are the flower pictures, produced by artists during the 17th, 18th and 19th centuries, which so accurately depict flowers in their incredible beauty. Scientists assert that there are over 270,000 species of flowers that have been documented and are existing in the 21st Century. The evolutionary history of flowers extends across some 125 years. During this time, an intricate assortment of more than 125,000 flower species has developed.
- During the Victorian era, in England the language of flowers was as important to people as being "well dressed." For example, the recognizable scent of a particular flower, plant or perhaps a scented handkerchief sent its own unique message. But scientists have yet to answer basic questions about these marvels of beauty. What led to their amazing diversity? Are there flowers that have not changed much during the evolution of this planet?
- Fossils of woody magnolia-like plants dating back 93 million years are the first evidence of plant life. More recently, tiny herb-like flower fossils dating back 120 million years have been uncovered by Paleobotanists have. Flowering plants, called angiosperms, were believed to be already diverse and found in most locations by the middle of the Cretaceous Period. 146 million years ago. A myriad of images of preserved flowers and flower parts have been found in fossils from Sweden, Portugal, England, and along the Eastern and Gulf coasts of the United States. Below are a few flowers which have a long history.
- The history of flowers is older than of humans. Flowers are ubiquitous. Virtually, every non-meat food that we eat starts as a flowering plant somewhere. Even the cotton clothes we wear come from flowering plants. Flowers appeared on earth about 130 million years ago, during the Crustacean period. Humans are believed to have existed for a mere two hundred thousand years. Once flowers took firm root about 100 million years ago, they quickly diversified into some 250,000 species.
- Flowering plants have the botanical name of 'angiosperm.' Angiosperms enclose their seed in fruit, and each fruit contains one or more carpels. Carpels are hollow chambers that protect and nourish the seeds. The

oldest fossil of an angiosperm was discovered in China in sediments that date back to 130 million years. It was so primitive that it had not developed the lovely flower that is typical of a flowering plant. It is still considered a flowering plant since it had carpels enclosing seeds that grew into fruits.

- ✰ The first angiosperms are believed to have evolved in areas where there were ecological disturbances like floodplains and volcanic regions. They were slow growing and had short life cycles. Thus, they matured and reproduced and multiplied much faster than trees. As they seeded quickly, they also evolved faster than other plants that were their competitors. Scientists believe that petals evolved 30 to 40 million years after the first angiosperms evolved. Once this happened, the angiosperms had the distinct advantage over all the other plants, as they attracted the birds and the bees, which, in turn, pollinated them.
- ✰ The world must have started bustling with activity with butterflies and bees and insects swarming everywhere. When this happened some 70 to 100 million years ago, the number of flowering species on Earth exploded. By the time the first flowering plant appeared, plant-eating dinosaurs had been around for a 100 million years. Dinosaurs were eating these and when the dinosaurs became extinct, another group of animals took their place–the mammals, which dispersed the angiosperm fruits, nuts and many vegetables. Now, the flower kingdom and the human race depended on each other.

Table 13.2: Name and History of the Flowers

Name of the Flower	*History of the Flower*
Alstroemeria	Alstroemeria is named after the Swedish botanist Baron Klas von Alstroemer. This South American flower's seeds were among many collected by von Alstroemer on a trip to Spain in 1753.
Aster	These plants were believed to have healing properties. Asters were laid on the graves of French soldiers to symbolize the wish that things turned out differently.
Calendula	The Romans used Calendula mixed with vinegar to season their meat and salad dishes. Calendula blossoms in wine were purported to soothe indigestion, and the petals as ointments, cured skin irritations, jaundice, sore eyes, and toothaches.
Carnations	These flowers were used in Greek ceremonial crowns. Carnation, comes from Greece. carnis(flesh) refers to the original color of the flower, or perhaps the word incarnacyon(incarnation), which refers to the incarnation of God made flesh.

Name of the Flower	*History of the Flower*
Chrysanthemums	Japanese emperors sat upon the Chrysanthemum throne. They put a single chrysanthemum petal on the bottom of a wine glass to sustain a long and healthy life. In Italy Chrysanthemums are associated with death.
Daisy	Beautiful gold hairpins, each ending in a daisy-like ornament were found when the Minoan palace on the Island of Crete was excavated. Daisy flowers are believed to be more than 4000 years old. Egyptian ceramics are also decorated with Daisies. "Marguerite", the French word for Daisy, is derived from a Greek word meaning "pearl". Francis I called his sister Marguerite of Marguerites and the lady used the Daisy as her device, so did Margaret of Anjou the wife of Henry IV and Margaret Beaufort, mother of Henry VII.
Dahlia	A herbal document written in Latin just sixty years after the coming of Columbus was discovered 1929. It noted that the Aztecs used dahlias as a treatment for epilepsy. Dahlias were late in coming to Europe.
	European scientific specialists considered the dahlia as a possible source of food since a disease had destroyed the French potato crop in the 1840s. Between 1800 and 1805, Lord and Lady Holland lived in France and in Spain where Lady Holland first saw dahlias that had been introduced to Spain about 15 years before. She sent some home to England and introduced the Dahlia into England.
Delphinium	Delphinium comes from the Greek word delphis, meaning dolphin - the flower resembles the bottle-like nose of a dolphin. Delphiniums were used by West Coast Native Americans to make blue dye, and European settlers made ink from ground delphinium flowers. The most ancient use of Delphinium flowers was a strong external concoction thought to drive away scorpions.
Gladiolus	The Latin word gladius, meaning "sword," and this flower was named for the shape of its leaves. Gladiolus was also called "xiphium," from the Greek word xiphos, also meaning sword. This flower is said to have represented the Roman gladiators. British Gladiolus used the stem base (corms) as a poultice and for drawing out thorns and splinters. In the 18th century, African Gladioli were imported in large quantities to Europe from South Africa.

Name of the Flower	*History of the Flower*
Holly	Medieval monks called this plant the Holy Tree. They believed Holly would keep evil spirits away, and protect their home from lightening. The early Romans decorated their hallways with garlands made from Holly for their mid-winter feast, Saturnalia. Later its pointed leaves represented the crown of thorns worn by Jesus, and the red berries his drops of blood.
Lily	Lilies have been associated with many ancient myths, and pictures of lilies were discovered in a villa in Crete, dating back to the Minoan Period, about 1580 B.C. Lilies are mentioned in the Old Testament, and in the New Testament, they symbolize chastity and virtue. In both the Christian and pagan traditions, the lily is a fertility symbol. In Greek marriage ceremonies the bride wears a crown of lilies and wheat implying purity and abundance.
Rose	The first cultivated roses appeared in Asian gardens more than 5,000 years ago. In ancient Mesopotamia, Sargon I, King of the Akkadians (2684-2630 B.C.) brought "vines, figs and rose trees" back from a military expedition beyond the River Tigris. Confucius wrote that during his life (551-479 B.C.), the Emperor of China owned over 600 books on the culture of Roses. Roses were introduced to Rome by the Greeks. During Roman public games all the streets were strewn with rose petals. Egyptian wall paintings depicting roses have been found in tombs dating from the fifth century B.C. to CleopatraÂ´s time. Cleopatra had a passion for everything Roman, and she is said to have scattered rose petals before Mark AnthonyÂ´s feet. Roses were introduced to Europe during the Roman Empire, where they were mainly used for ornamental purposes. Early Christians saw the rose as a symbol of paganism, orgy, and lust. King Childebert I had a rose garden planted for his Queen in Paris. Charlemagne ordered the cultivation of Roses. Leo IX, elected Pope in 1084, sent a Golden Rose to favored monarchs.

Name of the Flower	*History of the Flower*
Poinsettia	Dr. Joel Roberts-Poinsett, the US Ambassador to Mexico, brought the first poinsettia to the United States in 1928. Because Mexican legends say its bracts resemble the flower of Bethlehem, Poinsettias have the honor of decorating churches at Christmas time. Today, this flower is known worldwide as "the Christmas flower". This plant was used during the Medieval times as a purgative to rid the body of black bile and melancholy.
Queen Anne´s Lace	Queen AnneÂ´s Lace was named for Queen Anne, wife of King James I of England. The QueenÂ´s friends challenged her to create lace as beautiful as the flower. North African natives chewed it to protect themselves from the sun.
Snap Dragons	Snapdragons were common in the earliest gardens, but their actual origin is not known. Some botanists believe they grew wild in Spain and Italy.
Sun Flower	Sunflowers originated in Central and South America, and were grown for their usefulness, not their beauty. In 1532 Francisco Pizarro reported seeing the natives of the Inca Empire in Peru worshipping a giant sunflower. Incan priestesses wore large sunflower disks made of gold on their garments.
Tulip	Over a thousand years ago, Tulips grew wild in Persia, and near Kabul the Great Mogul Baber counted thirty-three different species. The word 'Tulip' is thought to be a corruption of the Turkish word for turbans. Persian poets sang its praises, and their artists drew and painted it so often, that all of Europe considered the Tulip to be the symbol of the Ottoman Empire. Wealthy people began to purchase tulip bulbs that were brought back from Turkey by Venetian merchants. In 1610, fashionable French ladies wore corsages of tulips, and many fabrics were decorated with tulip designs. In the seventeenth century, a small bed of tulips was valued at 15,000-20,000 francs. Tulipmania flourished between 1634-1637. just like the California Gold Rush, people abandoned jobs, businesses, wives, homes and lovers to become tulip growers. The frenzy spread from France, through Europe to the Low Countries.

Name of the Flower	*History of the Flower*
Violets	When Napoleon married Josephine, she wore Violets, and on each anniversary Josephine received a bouquet of violets. Following NapoleonÂ´s lead, the French Bonapartists chose the violet as their emblem, and nicknamed Napoleon "Corporal Violet". In 1814, Napoleon asked to visit Josephine's tomb before being exiled to the Island of St. Helena. When he died, he wore a locket around his neck that contained violets he had picked from JosephineÂ´s grave site.

Floriculture in India

Worldwide more than 140 countries are involved in commercial Floriculture. The leading flower producing country in the world is Netherlands and Germany is the biggest importer of flowers. Countries involved in the import of flowers are Netherlands, Germany, France, Italy and Japan while those involved in export are Colombia, Israel, Spain and Kenya. USA and Japan continue to be the highest consumers.

- About two decades back or so, the floriculture was just a pastime of rich people and hobby of flower lovers, but now it has opened a new vista in agri-business *i.e.,* commercial floriculture. With the increase in buying capacity of people, the flower lovers have now started buying them from the markets to beautify their home as well as to adore some one they love simply because they don't have time and enough space to grow flowers particularly in urban areas and in metropolitan cities. Flowers, it seems, is the most wanted item in any social occasions for conveying one's status and aesthetic sense. Flower is now so indispensable that one may cancel his her birthday celebration or Yama may postpone the death of a dying person in case flowers are not available at that time. No nuptial is performed and honeymoon of a young couple is not consummated till garden fresh rose and rajanigandha or tuberose with lingering and stupefying aroma are made available. Warm welcome cannot be offered to VIPs in the public functions without bouquet - flowers are so indispensable!
- All these, no doubt, have set flower business on a top gear. One may wonder, the global market on flower is at present, carrying a business worth 2000 crores US dollar (1992) par annum. India is also having a business worth '280 crores in her domestic market (1992–93).
- Indian Council of Agricultural Research (ICAR) conducted a survey of assessment on the possibilities of cut flowers trade in India during 1960–1962. An important conclusion was that an internal sale as '9.26

Crores worth flower weighing 10,460 tones grown in an area of 4000 hector. Flowers like Rose, Gladiolus, Tuberose, Chrysanthemum, Aster, Carnation, Orchids, Marigold are most popular in cut flower market all over the World.

Perfumes

- Perfumes have undergone through transformation stages for umpteenth times over since the beginning period till date. Now the times of designer perfumes have arrived. This has become after the fashion these days. Knowledgeable of the trade say, the chief ingredients of the fragrance are the scented oil.

Use of perfume is a must for those whose sweat smells foul. It has been from research that the sweats smell differently in men and women meaning thereby, men and women should make uses of different kinds of fragrances. The sweet of rose, juhi, lili are most suited to women while tobacco, mirchi, smells of junglee flowers are the best for men.

- India is the only country where the fragrances have been divided into 21 categories. The reason of this is there have been different customs and traditions of fragrances in our country. Their names have been given in accordance with the natures.
- The government of India had declared the year 1989 as the 'Fragrance Year'. There was a congregation of fragrance experts held here the same year round in which discussions were held on the topic of the 500 years old history of fragrance in India.

The Custom of Perfumes in the Olden Times

- The custom of fragrance remained well perched beginning right from the Sindhu Valley Civilization running through to the Gupta period up to the Maryann periods.
- The great poet Kalidas, in his work 'Abhigyan Shakuntalam' and 'meghdoot' makes details mentions of the use of fragrances by the heroin.
- The heroin of Kalids used to fragrant her hair and dresses using fragrances made up with sandalwood, jatamasi, kasturi etc. In the stone writings of the 10th decade are found the mentions of the descriptions of 'Gandh Kutirs' of Buddha. During the Kushan periods to, the mentions are found to have been made of gold embedded perfume boxes.
- Maha Kavi Vatsayan has also made mentions of the Indian tradition of fragrances. In his famous book 'Kama Sutra', Vatsayan said, "the females feel hypnotic attraction from the scent emanating from the male body and this smell creates attraction. This emanating smell of the male's body has of late been named as 'Sex Aroma'.

The Fashion of Perfume in Europe

- ✰ There has been a different custom of fragrances in Egypt. Under this custom, a special type of fragrance was dabbed on to the hair, the sweet fragrance of which wafted slowly filling into the air making the entire surrounding fragrant.
- ✰ It is also said in the eastern parts of African countries to Egypt, beautiful loban and ginger imported from India were used to prepare perfumes. The pharaoh of Egypt and his entire family members were always surrounded with fragrances to the extent that even in their coffins also, massive arrangements of perfumes used to be made. It is said that when the coffin of Tutankhamen was excavated on the surface 3000 years since it was engraved, scent were still wafting through his coffin.
- ✰ In the 7th decade, there was the tradition of making fragrances from the flowers in the graphi city of France. The beloved of the French king Louis 15th, Madam Pompidou had built a fragrance Bank. Probably this is why no fragrance can compete with the quality products of French fragrances.

Indian Tradition

- ✰ We find mentions of fragrance and perfume in the olden Indian epics and granths. The king began the day in the morning with fragrance. According to '*Agnipurana*', the kings took bathes with over 150 varieties of fragrances. The consumption of fragrances was enormous in the harems of kings. Force of men and women were engaged in preparing fragrances amongst whom the woman numbered the most. These women were called gandhkarika or gandhhadika. These fragrances were known as itra.
- ✰ In olden books, mentions are found of making itra with chandan and flowers. The Indian itras were exported to countries like Iran, Egypt, Arab, Turkey and Afghanistan. But so far as the itra was concerned, it had already reached there during the Vedic periods.

Itra in Mughals Period

- ✰ The intoxication for itra during the Moghal periods reached its acme. Mallika Noorjehan was a connoisseur of itras. She had prepared 'Itre Heena'. 5 sers of itras were prepared by mixing rose essences of 30 varieties of roses, 5 varieties of chandan oils, in which not even sufficient for Noorjehan even for a fortnight.
- ✰ Mughal had settled the Gandhies (experts in making itras) the kingdom, sanctioned them Zamindaries and let them make the different varieties of itras. Even to this day, the rose cultivation in a massive scale is going on in the Malwa borders of Rajasthan from where different species of beauteous roses are exported to the foreign countries and within the union territories.

The Quality of Perfumes

- ☆ A good quality perfume is the one which lasts longer and which reaches far and wide. Therefore the essential oils play significant role. To make a particular type of smell, a whole hog of labour is to be undertaken in preparing perfumes to attain a little quantity of the Ark. As for instance, from the orange flowers of 880 pounds, only 1 pound of essential oil is attained. From the 4000 pounds of rose petals is gotten I pound of rose oil. To preserve its fragrance for long, bio ingredients such as Mrig Kasturi, weaver Kasturi have to be mixed into it.
- ☆ Today, there is plethora of varieties of perfume available in the market in forms of liquid perfume, spray, cream, colon and oil. The spray colon is in fashion the most. Sprays smell longer. Besides, cream perfume in deodorant sticks is available which are antiseptic. As for that, deodorants are not longer lasting.

Medicinal Plants

- ☆ Medicinal plants (medicinal herbs) have been discovered and used in traditional medicine practices since pre-historic times. Plants synthesise hundreds of chemical compounds for functions including defence against insects, fungi, diseases, and herbivorous mammals. Numerous phytochemicals with potential or established biological activity have been identified. However, since a single plant contains widely diverse phytochemicals, the effects of using a whole plant as medicine are uncertain.
- ☆ The phytochemical content and pharmacological actions, if any, of many plants having medicinal potential remain unassessed by rigorous scientific research to define efficacy and safety. In the United States over the period 1999 to 2012, despite several hundred applications for new drug status, only two botanical drug candidates had sufficient evidence of medicinal value to be approved by the Food and Drug Administration.

According to the World Health Organization, most populations still rely on traditional medicines for their psychological and physical health requirements, since they cannot afford the products of Western pharmaceutical industries, together with their side effects and lack of healthcare facilities. Herbal medicines are in great demand in both developed and developing countries as a source of primary health care owing to their attributes having wide biological and medicinal activities, high safety margins and lesser costs. Herbal molecules are safe and would overcome the resistance produced by the pathogens as they exist in a combined form or in a pooled form of more than one molecule in the protoplasm of the plant cell. Researchers have identified number of compounds used in mainstream medicine which were derived from "ethnomedical" plant sources. Plants are used medicinally in different countries and are a source of many potent and powerful drugs.

☆ The earliest historical records of herbs are found from the Sumerian civilization, where hundreds of medicinal plants including opium are listed on clay tablets. The Ebers Papyrus from ancient Egypt describes over 850 plant medicines, while Dioscorides documented over 1000 recipes for medicines using over 600 medicinal plants in *De materia medica*, forming the basis of pharmacopoeias for some 1500 years. Drug research makes use of ethnobotany to search for pharmacologically active substances in nature, and has in this way discovered hundreds of useful compounds. These include the common drugs aspirin, digoxin, quinine, and opium. The compounds found in plants are of many kinds, but most are in four major biochemical classes: alkaloids, glycosides, polyphenols, and terpenes.

History

Prehistoric Times

☆ Plants, including many now used as culinary herbs and spices, have been used as medicines, not necessarily effectively, from prehistoric times.

☆ Spices have been used partly to counter food spoilage bacteria, especially in hot climates, and especially in meat dishes which spoil more readily. Angiosperms (flowering plants) were the original source of most plant medicines.

☆ Human settlements are often surrounded by weeds used as herbal medicines, such as nettle, dandelion and chickweed.

☆ Humans were not alone in using herbs as medicines: some animals such as non-human primates, monarch butterflies and sheep ingest medicinal plants when they are ill.

☆ Plant samples from prehistoric burial sites are among the lines of evidence that Paleolithic peoples had knowledge of herbal medicine.

☆ For instance, a 60 000-year-old Neanderthal burial site, "Shanidar IV", in northern Iraq has yielded large amounts of pollen from 8 plant species, 7 of which are used now as herbal remedies.

☆ A mushroom was found in the personal effects of *Ötzi the Iceman*, whose body was frozen in the Ötztal Alps for more than 5,000 years. The mushroom was probably used against whipworm.

Ancient Times

☆ In ancient Sumeria, hundreds of medicinal plants including myrrh and opium are listed on clay tablets. The ancient Egyptian Ebers Papyrus lists over 800 plant medicines such as aloe, cannabis, castor bean, garlic, juniper, and mandrake.

- From ancient times to the present, Ayurvedic medicine as documented in the Atharva Veda, the Rig Veda and the Sushruta Samhita has used hundreds of pharmacologically active herbs and spices such as turmeric, which contains curcumin.
- The Chinese pharmacopoeia, the *Shennong Ben Cao Jing* records plant medicines such as chaulmoografor leprosy, ephedra, and hemp. This was expanded in the Tang Dynasty *Yaoxing Lun*.
- In the fourth century BC, Aristotle's pupil Theophrastus wrote the first systematic botany text, *Historia plantarum*.
- In the first century AD, the Greek physician Pedanius Dioscorides documented over 1000 recipes for medicines using over 600 medicinal plants in *De materia medica*; it remained the authoritative reference on herbalism for over 1500 years, into the seventeenth century.

Middle Ages

- In the Early Middle Ages, Benedictine monasteries preserved medical knowledge in Europe, translating and copying classical texts and maintaining herb gardens. Hildegard of Bingen wrote *Causae et Curae* ("Causes and Cures") on medicine.
- In the Islamic Golden Age, scholars translated many classical Greek texts including Dioscorides into Arabic, adding their own commentaries. Herbalism flourished in the Islamic world, particularly in Baghdad and in Al-Andalus.
- Among many works on medicinal plants, Abulcasis (936–1013) of Cordoba wrote *The Book of Simples*, and Ibn al-Baitar (1197–1248) recorded hundreds of medicinal herbs such as *Aconitum*, nux vomica, and tamarind in his *Corpus of Simples*. Avicenna included many plants in his 1025 *The Canon of Medicine*. Abu-Rayhan Biruni, Ibn Zuhr, Peter of Spain, and John of St Amand wrote further pharmacopoeias.

Early Modern

- The Early Modern period saw the flourishing of illustrated herbals across Europe, starting with the 1526 *Grete Herball*. John Gerard wrote his famous *The Herball or General History of Plants* in 1597, based on Rembert Dodoens, and Nicholas Culpeper published his *The English Physician Enlarged*. Many new plant medicines arrived in Europe as products of Early Modern exploration and the resulting Columbian Exchange, in which livestock, crops and technologies were transferred between the Old World and the Americas in the 15th and 16th centuries. Medicinal herbs arriving in the Americas included garlic, ginger, and turmeric; coffee, tobacco and coca travelled in the other direction. In Mexico, the sixteenth century *Badianus Manuscript* described medicinal plants available in Central America.

19th and 20th Centuries

- The place of plants in medicine was radically altered in the 19th century by the application of chemical analysis. Alkaloids were isolated from a succession of medicinal plants, starting with morphine from the poppy in 1806, and soon followed by ipecacuanha and strychnos in 1817, quininefrom the cinchona tree, and then many others. As chemistry progressed, additional classes of pharmacologically active substances were discovered in medicinal plants. Commercial extraction of purified alkaloids including morphine from medicinal plants began at Merck in 1826.
- Synthesis of a substance first discovered in a medicinal plant began with salicylic acid in 1853. Around the end of the 19th century, the mood of pharmacy turned against medicinal plants, as enzymes often modified the active ingredients when whole plants were dried, and alkaloids and glycosides purified from plant material started to be preferred. Drug discovery from plants continued to be important through the 20th century and into the 21st, with important anti-cancer drugs from yew and Madagascar periwinkle.
- India has a rich culture of medicinal herbs and spices, which includes about more than 2000 species and has a vast geographical area with high potential abilities for Ayurvedic, Unani, Siddha traditional medicines but only very few have been studied chemically and pharmacologically for their potential medicinal value.
- Human beings have used plants for the treatment of diverse ailments for thousands of years. According to the World Health Organization, most populations still rely on traditional medicines for their psychological and physical health requirements, since they cannot afford the products of Western pharmaceutical industries, together with their side effects and lack of healthcare facilities.
- Rural areas of many developing countries still rely on traditional medicine for their primary health care needs and have found a place in day-to-day life. These medicines are relatively safer and cheaper than synthetic or modern medicine.
- People living in rural areas from their personal experience know that these traditional remedies are valuable source of natural products to maintain human health, but they may not understand the science behind these medicines, but knew that some medicinal plants are highly effective only when used at therapeutic doses.
- Herbal medicines are in great demand in both developed and developing countries as a source of primary health care owing to their attributes having wide biological and medicinal activities, high safety margins and lesser costs.

- Herbal molecules are safe and would overcome the resistance produced by the pathogens as they exist in a combined form or in a pooled form of more than one molecule in the protoplasm of the plant cell.
- With the advent of modern or allopathic medicine, it is noted that a number of important modern drugs have been derived from plants used by indigenous people. Traditional use of medicine is recognized as a way to learn about potential future medicines.
- Researchers have identified number of compounds used in mainstream medicine which were derived from "ethnomedical" plant sources. Plants are used medicinally in different countries and are a source of many potent and powerful drugs.

Chapter 14

Crop Significance and its Classification

Introduction

- Today, India ranks second worldwide in farm output. India is an agrarian country and more than 60 per cent of population depends on agriculture for their livelihood. While residing in urban areas may be we would not realize much importance of agriculture but this fact is not new that agriculture is the main source of income for major part of our country's population.

India's population is growing faster than its ability to produce rice and wheat. The required level of investment for the development of marketing, storage and cold storage infrastructure is estimated to be huge. The country produces innumerable crops ranging from medicinal to cereal crops. These commodities are used for various purposes from human consumption, in industries, for animal feed etc.

- Most of us do not know major categories of crops. We are unaware of its valuable contribution to our economy. Ignoring the negative approach of the citizens living in urban areas, the cropping activities are continue to go on all the year-round in India, provided water is available for crops.

Classification of Crops

- In India crops are broadly classified into three major categories, *viz.*
  - *Kharif crops:* The crops that are sown in the rainy season are called kharif crops. Its season starts from July and ends in October, *e.g.* maize,

sugarcane, cotton, jawar, bajra, soyabean, turmeric, paddy, moong, ground nuts, red chillies.

- *Rabi crops:* The crops that are sown in the winter season are called rabi crops. Its season is during October to March, *e.g.,* wheat, oil seed, pulses, rubber beans.
- *Zaid crops:* Crops grown between March and June are known as Zaid *e.g.* muskmelon, watermelon and vegetables like guard, pumpkin *etc.*

☆ At some places rice is produced more in quantity and at some other places more wheat is produced. In some regions maize, jute and in other regions sugar-cane is produced. India produces different types of crops due to difference in soil and climate.

Main Crops

(i) Food Crops

- *Rice:* Rice is the main grain crop of India. India ranks second in the world in production of rice. About 34 per cent of the total cultivated area of the nation is under rice cultivation. Rice is cultivated in areas having annual average rainfall of 125 cm. Major rice cultivating areas are north east India, eastern and western coastal regions. West Bengal, Punjab and Uttar Pradesh are the major rice producing states.
- *Wheat:* Wheat is the second major crop in India. Wheat is cultivated in areas with mean annual rainfall of 75 cm and fertile soil. Wheat has got an important role in 'Green Revolution'. The highest quantity of wheat in the country is in Uttar Pradesh. 35 per cent of wheat is produced only in Uttar Pradesh. This is produced by Punjab and Haryana where production of wheat is on a large scale.
- *Maize:* Maize is an important khaki crop of rainy season. Maize is cultivated in different areas and in different climates but it is suitable where temperature is 35° Celsius and rainfall is 75 centimeters. It is cultivated in hilly areas-of Jammu and Kashmir and Himachal Pradesh. Maize is cultivated throughout our country but it is cultivated more in Punjab, U.P., Bihar, M.P. and Rajasthan.
- *Pulses:* Pulses are grown in dry climate region in India. These crops provide nitrogen to the soil. Pulses are a source of proteins in the diet. Madhya Pradesh is the leading pulses producing state in India. It is also produce in Uttar Pradesh and Rajasthan.
- *Jowar:* This crop is grown where the climate is hot and dry. It is cultivated in Maharashtra, Karnataka, Andhra Pradesh, and Tamil Nadu.

☆ In agriculture there are few crops which are grown for profit are called as 'Cash Crops or Commercial crops'. Cash crop is a backbone of agriculture economy of India. It sets a strong base for Indian economy where country's trade and commerce flourish domestically and internationally. Cash crops are generally grown for money. In earlier days, cash crops were grown in a very small scale but today it forms a major contribution to our nation's economy. Now it has grown at large scale for commercial purpose.

(ii) Cash Crops

- *Sugarcane:* Sugarcane is an important cash crop of India. Molasses, sugar and khandasari *etc.* are produced from the juice of sugarcane. Sugarcane cultivation needs temperature of 15° to 40° and rainfall of 100 to 150 centimeters and fertile loamy soil or hard soil. Sugarcane is cultivated from Kanyakumuri (southern part) to Punjab (north-west) but it is more cultivated in Uttar Pradesh.
- *Coffee:* There is a great demand of coffee in the world market. For reason, India exports coffee. Coffee cultivation needs hot and wet climate and fertile sloppy land. It is mainly produced in south Indian states of Karnataka, Kerala and Tamil Nadu. Coffee is produced on a large scale on the mountain ranges of Nilgiri.
- *Tea:* India is first in the cultivation of tea in the world. Tea cultivation needs hot climate, excess rainfall and sloppy soil. Tea is found more in Assam, Karnataka, Kerala and Himachal Pradesh, Dehradun, Ranchi and Tripura.
- *Rubber:* Rubber is needed by different industries and transport industry in this modern age. It is cultivated in hot and wet climate garden in natural way. It is cultivated in the State of Kerala in India. Except Kerala, Andaman Nicobar Islands, Kurgan of Karnataka State and Chicmagalur district *etc.*
- *Oilseeds:* Groundnut, mustard, rapeseed, linseed and caster help us to get our edible oil. Oil is also extracted from coconut. India occupies the first position in the world in the production of groundnut. Groundnut cultivation needs temperature varying from 20° to 30° degree Celsius and needs 60 to 80 centimeters of rainfall. Groundnut cultivation needs sandy and light soil. Groundnuts are produced more in Tamil Nadu, Maharashtra, Gujarat, Andhra Pradesh and Uttar Pradesh.

(iii) Fibre Crops

- *Cotton:* Cultivation of cotton needs 20° to 25° Celsius temperature and 50 to 75 centimeters of rainfall. Cotton plant needs wet climate at the time of growing and dry climate at the time of collecting seeds. it is known as Kharif crop. Cotton cultivation takes place in Gujarat, Maharashtra, Punjab and Haryana of our country. Except these States, cotton cultivation also takes place in Karnataka, Tamilnadu, M.P, Rajasthan, A.P. and U.P.

❀ *Jute:* India is a major producer of jute in the world. It is another type of fiber crop. Bags, ropes and a lot of other things are made out of jute. Jute cultivation needs hot and wet climate and fertile loamy land. Temperature of 24° to 35° Celsius and 90 to 150 centimeters of rainfall is suitable for jute cultivation. In our country, jute is cultivated in Odhisa, West Bengal, eastern U.P., Bihar, Assam and Tripura.

Table 14.1: Important World Food Crops

Crop	*Botanical Name*	*Family*	*Annual or Perennial*
Wheat	*Triticum aestivum*	Poaceae†	Annual
Rice, Paddy	*Oryza sativa*	Poaceae	Annual
Maize	*Zea mays*	Poaceae	Annual
Soybeans	*Glycine max*	Fabaceae	Annual
Barley	*Hordeum vulgare*	Poaceae	Annual
Sorghum	*Sorghum bicolor*	Poaceae	Annual
Millet	*Setaria, Echinochloa, Eleusine, Panicum, Pennisetum* spp.	Poaceae	Annual
Groundnuts in Shell (peanuts)	*Arachis hypogaea*	Fabaceae	Annual
Beans, Dry	*Phaseolus* spp.	Fabaceae	Annual
Rapeseed	*Brassica napus*	Brassicaceae	Annual
Sugar Cane	*Saccharum officinarum*	Poaceae	Perennial
Sunflower Seed	*Helianthus annuus*	Asteraceae	Annual
Potatoes	*Solanum tuberosum*	Solanaceae	Annual
Cassava	*Manihot esculenta*	Euphorbiaceae	Perennial
Oats	*Avena sativa*	Poaceae	Annual
Coconuts	*Cocos nucifera*	Arecaceae	Perennial
Oil Palm Fruit	*Elaeis guineensis*	Arecaceae	Perennial
Chick-Peas	*Cicer arietinum*	Fabaceae	Annual
Coffee, Green	*Coffea* spp.	Rubiaceae	Perennial
Rye	*Secale cereale*	Poaceae	Annual
Sweet Potatoes	*Ipomoea batatas*	Convolvulaceae	Annual
Cowpeas, Dry	*Vigna unguiculata*	Fabaceae	Annual
Olives	*Olea europaea*	Oleaceae	Perennial
Grapes	*Vitis vinifera*	Vitaceae	Perennial
Sesame Seed	*Sesamum indicum*	Pedaliaceae	Annual
Cocoa Beans	*Theobroma cacao*	Sterculiaceae	Perennial
Sugar Beets	*Beta vulgaris*	Chenopodiaceae	Annual

Crop	*Botanical Name*	*Family*	*Annual or Perennial*
Peas, Dry	*Pisum sativum*	Fabaceae	Annual
Apples	*Malus pumila*	Rosaceae	Perennial
Plantains	*Musa* spp.	Musaceae	Perennial
Bananas	*Musa* spp.	Musaceae	Perennial
Yams	*Dioscorea* spp.	Dioscoreaceae	Annual
Tomatoes	*Lycopersicon esculentum*	Solanaceae	Annual

Source: Production figures from FAOSTAT Agriculture data bases for 2002.

†Poaceae=Graminae; Fabaceae=Leguminosae; Brassicaceae=Cruciferae; Asteraceae=Compositae; Aracaceae=Palmae

Types of Food Crops and their Significance

Cereals

- ☆ Cereals are edible seeds from the grass family. They include wheat, rice, corn, barley, oats, sorghum, and millet, among others. The grain is a caryopsis, which is a dry, one-seeded fruit, with a hard-outer pericarp fused to the seed coat. The grains of different cereal crops may differ in size and shape, but they all have a similar structure and composition.
- ☆ The endosperm is predominantly starch. The aleurone layer and embryo (germ) are rich in protein, vitamins and minerals, but are often removed during processing. 100 μg of whole grain provides about 350 kcal, 8 to 12 μg of protein and useful amounts of calcium, iron (though phytic acid may hinder absorption) and the B vitamins. They lack vitamin C, and with the exception of yellow maize, also lack carotene.
- ☆ For a healthy diet, they should be consumed with other foods rich in vitamins A and C and minerals. The protein of some cereals is deficient in the essential amino acids lysine and tryptophan, and should be supplemented with other sources of protein such as legumes or animal products.

Pulses

- ☆ They are the edible seeds of members of the Fabaceae family, which includes beans, peas, soybeans, groundnuts (peanuts) and lentils, among other crops. They provide an important source of protein and B vitamins in the diet, as well as carbohydrate. Protein quality is not quite as good as in meat, because legumes generally lack adequate levels of the essential amino acid methionine. However, this limitation can be overcome by consuming pulses and cereals together.

Roots and Tubers

- Edible tubers, roots and corms are widely consumed throughout the world. In the tropics, cassava, sweet potatoes, taro, yams and arrowroot are staple food crops. Potato is widely grown in temperate and subtropical climates.
- The yield potential of these crops is very high. However, they are generally low in protein, minerals and vitamins in comparison to cereal crops. There is some variation among species in this regard - taro and yams have up to 6 per cent good quality protein, and potatoes provide some minerals and vitamin C. It is expected that roots and tubers will play an increasingly important role as food security crops.

Oilseed Crops

- Important oil seed crops include soybeans, rapeseed, sunflower seeds, groundnuts, oil palm, sesame, and cottonseed. Some oil crops are important locally, such as Shea butter in West Africa.

Vegetable Crops

- The foods called vegetables are a diverse group that include some fruits (*e.g.*, tomatoes), roots (*e.g.*, carrots), and flowers (*e.g., broccoli*). Nonetheless, "vegetable" is a useful term that refers to crops that are generally eaten fresh or preserved in the fresh state.
- Vegetables are a very important part of the diet. They are nearly all rich in carotene and vitamin C and contain significant amounts of calcium, iron and other minerals. They are not usually good sources of B vitamins, energy, and protein. They often contain high amounts of dietary fibre. When looking at world food production figures it is easy to underestimate the importance of vegetables in the diet, because many species are grown and consumed locally, and so do not make the list of 30-40 most widely grown crops.
- Dark green vegetables such as amaranth and cassava leaves typically eaten in tropical countries are far superior nutritionally to cabbage and lettuce. An increase in consumption of these crops could help to ameliorate the problem of vitamin A deficiency that is widespread in developing countries.

Fruits

- A wide variety of fruits grow wild or are cultivated throughout the world. Fruits are often good sources of vitamin C, and many contain useful quantities of carotene. Fruits usually contain very little fat or protein and little or no starch. The carbohydrate is present in the form of various sugars.

Table 13.2: Direct Contribution of Selected Food Crops to per Capita Consumption of Calories, Protein and Fat

Crop	*Developed Countries*			*Developing Countries*		
	Cal/Cap/ Day	*Prot/Cap/ Day (g)*	*Fat/Cap/ Day (g)*	*Cal/Cap/ Day*	*Prot/Cap/ Day (g)*	*Fat/Cap/ Day (g)*
Wheat	**739.0**	**23.1**	2.9	478.3	**13.8**	2.0
Rice (Paddy Equivalent)	117.9	2.2	0.2	**703.2**	**13.1**	1.7
Maize	89.6	1.9	0.5	175.9	4.3	1.5
Maize Germ Oil	14.4	0.0	1.6	3.7	0.0	0.4
Soyabeans	9.8	0.9	0.4	26.1	2.5	0.9
Soyabean Oil	153.2	0.1	**17.3**	49.6	0.0	**5.6**
Groundnuts (in Shell Eq)	16.6	0.8	1.4	23.2	1.0	1.9
Groundnut Oil	5.8	0.0	0.7	22.4	0.0	2.5
Beans	13.2	0.9	0.1	22.8	1.4	0.1
Potatoes	**132.0**	3.2	0.2	40.4	1.0	0.1
Cassava	0.1	0.0	0.0	**55.7**	0.4	0.1
Sweet Potatoes	2.9	0.0	0.0	37.8	0.4	0.1
Tomatoes	13.4	0.6	0.2	5.9	0.3	0.1
Rape and Mustard Oil	75.5	0.0	**8.5**	25.1	0.0	2.8
Sunflower Oil	88.2	0.0	**10.0**	19.2	0.0	2.2
Palm Oil	12.4	0.0	1.4	**50.0**	0.0	**5.7**

Source: FAOSTAT Nutrition Databases, 2000.

☆ Fruits are high in cellulose, which contributes to dietary fibre. The citrus fruits, such as oranges, lemons, grapefruits, tangerines and limes, contain good quantities of vitamin C but little carotene. Papayas and mangoes contain both carotene and vitamin C.

Nuts

☆ Important crops in this category include coconut, cashew nut, almond, and walnut, among others. These crops tend to be high in fat, but may also contain beneficial vitamins, minerals, and protein.

Spices

☆ Aside from our need for salt, few of these flavorings have much nutritional importance, but all serve to make the food more pleasing to the taste.

Beverage Crops

- This group includes a diversity of drinks from plant sources, including fruit juices, tea, coffee, beer, wine, and spirits. The main contribution of this group to human nutrition is to supply water, essential for human life.
- Some also provide vitamins and minerals. Others provide stimulants (caffeine) or alcohol for relaxation.

Significance and Contribution to Nutrition

- Cereal crops provide by far the most calories in the diet. Wheat is the leading crop for direct human consumption in developed countries, whereas rice is the leading crop in developing countries.
- Although the pulse crops have a higher protein content than cereal crops, cereals are the most important source of protein, because they are consumed in much greater quantities.
- Soybean is the most important oil crop in developed countries, followed by rapeseed and sunflower oil. In developing countries, soybean and palm oil are the leading oil crops.
- Potato is the predominant starchy root and tuber crop in developed countries. Cassava is the lead crop in this category in developing countries, but potatoes and sweet potatoes are also important.
- People in developing countries obtain more calories from cereal crops than in the developed world. Calories obtained from oil crops are relatively high in developed countries.

Chapter 15

Importance, Scope, Researches and National Agricultural Setup

Introduction

- ✰ Agriculture is a very broad term encompassing all aspects of crop production, livestock farming, fisheries, forestry *etc*. The scenario in food production in changing fast in the country with the advances made in all branches of Agricultural Sciences. However, the science of Agronomy, a specialized subject dealing with all aspects of field crop production, has accelerated the pace of food production, aided by the progress made in understanding the intricate relationships between crop growth and yield, and between crop and its environment like climate, soil, biotic factors and management practices.

Definitions

- ✰ Agriculture is defined in the Agriculture Act 1947, as including 'horticulture, fruit growing, seed growing, dairy farming and livestock breeding and keeping, the use of land as grazing land, meadow land, osier land, market gardens and nursery grounds, and the use of land for woodlands where that use ancillary to the farming of land for Agricultural purposes".
- ✰ It is also defined as 'purposeful work through which elements in nature are harnessed to produce plants and animals to meet the human needs. It is a biological production process, which depends on the growth and development of selected plants and animals within the local environment.

Agriculture as Art, Science and Business of Crop Production

- Agriculture is defined as the art, the science and the business of producing crops and the livestock for economic purposes.
- As an art, it embraces knowledge of the way to perform the operations of the farm in a skillful manner. The skill is categorized as;
    - *Physical Skill:* It involves the ability and capacity to carry out the operation in an efficient way for *e.g.,* handling of farm implements, animals *etc.,* sowing of seeds, fertilizer and pesticides application *etc.*
    - *Mental Skill:* The farmer is able to take a decision based on experience, such as *(i)* time and method of ploughing, *(ii)* selection of crop and cropping system to suit soil and climate, *(iii)* adopting improved farm practices *etc.*
- *As a Science:* It utilizes all modern technologies developed on scientific principles such as crop improvement/breeding, crop production, crop protection, economics *etc.,* to maximize the yield and profit. For example, new crops and varieties developed by hybridization, transgenic crop varieties resistant to pests and diseases, hybrids in each crop, high fertilizer responsive varieties, water management, herbicides to control weeds, use of bio-control agents to combat pest and diseases *etc.*
- *As the Business :* As long as agriculture is the way of life of the rural population, production is ultimately bound to consumption. But agriculture as a business aims at maximum net return through the management of land, labour, water and capital, employing the knowledge of various sciences for production of food, feed, fibre and fuel. Now agriculture is commercialized to run as a business through mechanization.
- Agriculture has no single and simple origin. It was started in different part of the world during different periods. Present day agriculture in India as elsewhere has evolved itself through the ages. As India was a pastoral country before agriculture was started, development of crops and animal took place concurrently leading to various types of farming systems that are now seen in different parts of the country.
- India's most important contribution to world agriculture is rice, the staple food crop of most of south, south-east and East Asia. Sugarcane, legumes and tropical fruit like mango are also natives of India. Scientific agriculture began in India when sugarcane, cotton and tobacco were grown for export purposes. In 1870, a joint department of agriculture, revenue and commerce were established. Later, on the recommendation of the Famine Commission of 1880, a separate department was started with the objective of increasing food production for local people and industrial raw materials for export.

Scope of Agriculture

- Proverbially, India is known as *"Land of Villages"*. Near about 67 per cent of India's population live in villages. The occupation of villagers is agriculture. Agriculture is the dominant sector of our economy and contributes in various ways such as:
  - *National Economy:* In 1990–1991, agriculture contributed 31.6 per cent of the National Income of India, while manufacturing sector contributed 17.6 per cent. It is substantial than other countries for example in 1982 it was 34.9 per cent in India against 2 per cent in UK, 3 per cent in USA, 4 per cent in the Canada. It indicated that the more the more the advanced stage of development the smaller is the share of agriculture in National Income.
  - *Total Employment:* Around 65 per cent population is working and depends on agriculture and allied activities. Nearly 70 per cent of the rural population earns its livelihood from agriculture and other occupation allied to agriculture. In cities also, a considerable part of labor force is engaged in jobs depending on processing and marketing of agricultural products.
  - *Industrial Inputs:* Most of the industries depend on the raw material produced by agriculture, so agriculture is the principal source of raw material to the industries. The industries like cotton textile, jute, paper, sugar depends totally on agriculture for the supply of raw material. The small scale and cottage industries like handloom and power loon, ginning and pressing, oil crushing, rice husking, sericulture fruit processing, *etc.* are also mainly agro-based industries.
  - *Food Supply:* During this year targeted food production was 198 million tons and which is to be increased 225 million tons by the end of this century to feed the growing population of India *i.e.,* 35 corer in 1951 and 100 corers at the end of this century. India, thus, is able to meet almost all the need of its population with regards to food by develop intensive programme for increasing food production.
  - *State Revenue:* The agriculture is contributing the revenue by agriculture taxation includes direct tax and indirect tax. Direct tax includes land revenue, cesses and surcharge on land revenue, cesses on crops and agricultural income tax. Indirect tax induces sales tax, custom duty and local octri, *etc.* which farmer pay on purchase of agriculture inputs.
  - *Trade:* Agriculture plays and important role in foreign trade attracting valuable foreign exchange, necessary for our economic development. The product from agriculture based industries such as jute, cloth, tinned food, *etc.* contributed to 20 per cent of our export. Around

50 per cent of total exports are contributed by agril-sector. Indian agriculture plays and important role in roads, rails and waterways outside the countries. Indian in roads, rails and waterways used to transport considerable amount of agril-produce and agro-based industrial products. Agril-products like tea, coffee, sugar, oil seeds, tobacco; spices, *etc.* also constitute the main items of export from India.

Importance of Agriculture in India

- With a 16 per cent contribution to the gross domestic product (GDP), agriculture still provides livelihood support to about two-thirds of country's population.
- The sector provides employment to 58 per cent of country's work force and is the single largest private sector occupation.
- Agriculture accounts for about 15 per cent of the total export earnings and provides raw material to a large number of Industries (textiles, silk, sugar, rice, flour mills, milk products).
- Rural areas are the biggest markets for low-priced and middle-priced consumer goods, including consumer durables and rural domestic savings are an important source of resource mobilization.
- The agriculture sector acts as a wall in maintaining food security and in the process, national security as well.
- The allied sectors like horticulture, animal husbandry, dairy and fisheries, have an important role in improving the overall economic conditions and health and nutrition of the rural masses.
- To maintain the ecological balance, there is need for sustainable and balanced development of agriculture and allied sectors.
- Agriculture's eyes and minds are soothed by dynamic changes from brown (bare soil) to green (growing crop) to golden (mature crop) and bumper harvests.
- Plateauing of agricultural productivity in irrigated areas and in some cases the declining trend warrants attention of scientists. Agriculture helps to elevate the community consisting of different castes and communities to a better social, cultural, political and economical life. Agriculture maintains a biological equilibrium in nature. Satisfactory agricultural production brings peace, prosperity, harmony, health and wealth to individuals of a nation by driving away distrust, discord and anarchy.

Revolution in Agriculture

- Green revolution: Food grain production
- White revolution: Milk production

- ☆ Yellow revolution: Oilseeds production
- ☆ Gray revolution: Manures and Fertilizers
- ☆ Blue revolution: Fish production
- ☆ Red revolution: Meat/Tomato production
- ☆ Round revolution: Potato production
- ☆ Silver revolution: Egg production/Poultry
- ☆ Pink revolution: Prawn production
- ☆ Golden revolution: Fruit production (apple)
- ☆ Golden fibre revolution: Jute Production
- ☆ Brown revolution: Non-conventional energy source
- ☆ Black revolution: Bio-fuel (Jatropha) production
- ☆ Rainbow revolution: Overall development of agriculture sector (1996)
- ☆ Food chain revolution: Food grain production
- ☆ Evergreen revolution: Reduction in wastage of food grains, fruits and vegetables
- ☆ Second Green revolution: Protein rich pulses

PM Modi's Tricolour Revolution

- ☆ White revolution: Cattle welfare
- ☆ Blue revolution: For fishermen's welfare and clean water
- ☆ Saffron revolution: Solar energy

Branches of Agriculture

- ☆ *Agronomy:* Deals with the production of various crops which includes food crops, fodder crops, fibre crops, sugar, oilseeds, *etc.* The aim is to have better food production and how to control the diseases.
- ☆ *Horticulture:* Deals with the production of fruits, vegetables, flowers, ornamental plants, spices, condiments and beverages.
- ☆ *Forestry:* Deals with production of large scale cultivation of perennial trees for supplying wood, timber, rubber, *etc.* and also raw materials for industries.
- ☆ *Animal Husbandry:* Deals with agricultural practice of breeding and raising livestock in order to provide food for humans and to provide power (drought) and manure for crops.
- ☆ *Fishery science:* Deals with practice of breeding and rearing fishes including marine and inland fishes, shrimps, prawns *etc.* in order to provide food, feed and manure.
- ☆ *Agricultural Engineering:* Deals with farm machinery for filed preparation, inter-cultivation, harvesting and post harvest processing including soil and water conservation engineering and bio-energy.

- *Home Science:* Deals with application and utilization of agricultural produces in a better manner in order to provide nutritional security, including value addition and food preparation.

Evolution of Man and Agriculture

There are different stages in development of agriculture, which is oriented with human civilization. They are Hunting Pastoral Crop culture Trade (stages of human civilization).

- *Hunting:* It was the primary source of food in old days. It is the important occupation and it existed for a very long period.
- *Pastoral:* Human obtained his food through domestication animals, *e.g.* dogs, horse, cow, buffalo, *etc.* They lived in the periphery of the forest and they had to feed his domesticated animals. For feeding his animals, he would have migrated from one place to another in search of food. It was not comfortable and they might have enjoyed the benefit of staying in one place near the river bed.
- *Crop culture:* By living near the river bed, he had enough water for his animals and domesticated crops and started cultivation. Thus, he has started to settle in a place.
- *Trade:* When he started producing more than his requirement the excess was exchanged, this is the basis for trade. When agriculture has flourished, trade developed. This lead to infrastructure development like road, routes, *etc.* Agriculture became civilized from crop culture stage.

National Agricultural Setup in India

- The highest body controlling agricultural research and education in India is" Indian Council of Agricultural Research (ICAR)." It was established on July 16, 1929 with the name "Imperial Council of Agricultural Research" under the Societies Registration Act, 1860 in pursuance of the report of the Royal Commission on Agriculture. ICAR headquarters at Krishi Bhavan, New Delhi.
- Union Minister of Agriculture is the ex-officio President of the ICAR Society.
- Secretary, Department of Agricultural Research and Education Ministry of Agriculture, Govt. of India and Director-General, ICAR is the Principal Executive Officer and chief administrative authority of Council ICAR.
- Agricultural Scientists' Recruitment Board, Chairman.
- Directorate of Information and Publications of Agriculture, New Delhi.
- Deputy Directors-General (8)
- Additional Secretary (DARE) and Secretary (ICAR).

- ☆ Additional Secretary and Financial Advisor.
- ☆ Assistant Directors-General (24).
- ☆ Directorates/Project Directorates - 25 (with upgradation of 12 NRCs).
- ☆ National Bureaux – 6 (New-NBAII, Bengaluru and NBAIM, Mau, UP).
- ☆ Deemed Universities status -6 (New- NAARM, Hyderabad and NIASM, Malegaon, Maharashtra).
- ☆ National Bureau of Agriculturally Important Insects (NBAII) [formerly Project Directorate of Biological Control (PDBC)] is a nodal Institute at national level for research and development on all aspects of work on harnessing resources of insects, including biological control of crop pests and weeds, training, information repository, technology dissemination and national/international cooperation (2009).
- ☆ National Bureau of Agriculturally Important Microorganism (NBAIM), Mau, UP (2005).
- ☆ There are 44 Agricultural Technology Information Centres (ATIC) established under ICAR institutes.
- ☆ ICAR Introduced revised curricula and syllabi for 95 disciplines in Master's and 80 disciplines in Doctoral programmes.

Objectives of the ICAR

1. To encourage and co-ordinate education and research in Agriculture, Animal Husbandry and Fishery and to help in utilization of result of research.
2. To act as a centre of distribution for researches related to agriculture and Animal Science and general information.
3. Establishment and maintenance of research and contact library.
4. To do all that is necessary for fulfilment of the above stated objectives.
5. To provide advisory service in education, research and training in Agriculture and related fields of science.

List of ICAR Institutes

1. ICAR-Central Island Agricultural Research Institute, Port Blair
2. ICAR-Central Arid Zone Research Institute, Jodhpur
3. ICAR-Central Avian Research Institute, Izatnagar
4. ICAR-Central Inland Fisheries Research Institute, Barrackpore
5. ICAR-Central Institute Brackishwater Aquaculture, Chennai
6. ICAR-Central Institute for Research on Buffaloes, Hissar
7. ICAR-Central Institute for Research on Goats, Makhdoom
8. ICAR-Central Institute of Agricultural Engineering, Bhopal

9. ICAR-Central Institute for Arid Horticulture, Bikaner
10. ICAR-Central Institute of Cotton Research, Nagpur
11. ICAR-Central Institute of Fisheries Technology, Cochin
12. ICAR-Central Institute of Freshwater Aquaculture, Bhubneshwar
13. ICAR-Central Institute of Research on Cotton Technology, Mumbai
14. ICAR-Central Institute of Sub Tropical Horticulture, Lucknow
15. ICAR-Central Institute of Temperate Horticulture, Srinagar
16. ICAR-Central Institute on Postharvest Engineering and Technology, Ludhiana
17. ICAR-Central Marine Fisheries Research Institute, Kochi
18. ICAR-Central Plantation Crops Research Institute, Kasargod
19. ICAR-Central Potato Research Institute, Shimla
20. ICAR-Central Research Institute for Jute and Allied Fibres, Barrackpore
21. ICAR-Central Research Institute of Dryland Agriculture, Hyderabad
22. ICAR-National Rice Research Institute, Cuttack
23. ICAR-Central Sheep and Wool Research Institute, Avikanagar, Rajasthan
24. ICAR- Indian Institute of Soil and Water Conservation, Dehradun
25. ICAR-Central Soil Salinity Research Institute, Karnal
26. ICAR-Central Tobacco Research Institute, Rajahmundry
27. ICAR-Central Tuber Crops Research Institute, Trivandrum
28. ICAR-ICAR Research Complex for Eastern Region, Patna
29. ICAR-ICAR Research Complex for NEH Region, Barapani
30. ICAR-Central Coastal Agricultural Research Institute, Ela, Old Goa, Goa
31. ICAR-Indian Agricultural Statistics Research Institute, New Delhi
32. ICAR-Indian Grassland and Fodder Research Institute, Jhansi
33. ICAR-Indian Institute of Agricultural Biotechnology, Ranchi
34. ICAR-Indian Institute of Horticultural Research, Bengaluru
35. ICAR-Indian Institute of Natural Resins and Gums, Ranchi
36. ICAR-Indian Institute of Pulses Research, Kanpur
37. ICAR-Indian Institute of Soil Sciences, Bhopal
38. ICAR-Indian Institute of Spices Research, Calicut
39. ICAR-Indian Institute of Sugarcane Research, Lucknow
40. ICAR-Indian Institute of Vegetable Research, Varanasi
41. ICAR-National Academy of Agricultural Research and Management, Hyderabad
42. ICAR-National Institute of Biotic Stresses Management, Raipur

43. ICAR-National Institue of Abiotic Stress Management, Malegaon, Maharashtra
44. ICAR-National Institute of Animal Nutrition and Physiology, Bengaluru
45. ICAR-National Institute of Research on Jute and Allied Fibre Technology, Kolkata
46. ICAR-National Institute of Veterinary Epidemiology and Disease Informatics, Hebbal, Bengaluru
47. ICAR-Sugarcane Breeding Institute, Coimbatore
48. ICAR-Vivekananda Parvatiya Krishi Anusandhan Sansthan, Almora
49. ICAR-Central Institute for Research on Cattle, Meerut, Uttar Pradesh
50. ICAR-National Institute of High Security Animal Diseases, Bhopal
51. ICAR-Indian Institute of Maize Research, New Delhi
52. ICAR- Central Agroforestry Research Institute, Jhansi
53. ICAR-National Institute of Agricultural Economics and Policy Research, New Delhi
54. ICAR- Indian Institute of Wheat and Barley Research, Karnal
55. ICAR- Indian Institute of Farming Systems Research, Modipuram
56. ICAR- Indian Institute of Millets Research, Hyderabad
57. ICAR- Indian Institute of Oilseeds Research, Hyderabad
58. ICAR- Indian Institute of Oil Palm Research, Pedavegi, West Godawari
59. ICAR- Indian Institute of Water Management, Bhubaneshwar
60. ICAR-Indian Institute of Rice Research, Hyderabad
61. ICAR- Central Institute for Women in Agriculture, Bhubaneshwar
62. ICAR-Central Citrus Research Institute, Nagpur
63. ICAR-Indian Institute of Seed Research, Mau
64. ICAR-Indian Agricultural Research Institute, Jharkhand.

Deemed Universities

1. ICAR-Indian Agricultural Research Institute, New Delhi
2. ICAR-National Dairy Research Institute, Karnal
3. ICAR-Indian Veterinary Research Institute, Izatnagar
4. ICAR-Central Institute on Fisheries Education, Mumbai

Universities with Agricultural Faculty

Andhra Pradesh

1. Acharya NG Ranga Agricultural University, Guntur
2. Dr. YSRHU (APHU), Venkataramannagudem

3. Sri Venkateswara Veterinary University, Tirupati

Assam

4. Assam Agricultural University, Jorhat

Bihar

5. Bihar Agricultural University, Sabour, Bhagalpur
6. Bihar Animal Sciences University, Patna

Chhattisgarh

7. Indira Gandhi Krishi Viswa Vidhyalaya, Raipur
8. Chhattisgarh Kamdhenu Visvavidyalaya, Durg

Gujarat

9. Sardar Krushinagar Dantiwada Agricultural University, Dantiwada
10. Anand Agricultural University, Anand
11. Navsari Agricultural University, Navsari
12. Junagarh Agricultural University, Junagarh
13. Kamdhenu University, Gandhinagar

Haryana

14. Chaudhary Charan Singh Haryana Agricultural University, Hisar
15. Lala Lajpat Rai University of Veterinary and Animal Sciences, Hisar
16. Haryana State University of Horticultural Sciences, Karnal

Himachal Pradesh

17. Ch. Sarwan Kumar Himachal Pradesh Krishi Viswavidyalaya, Palampur
18. Dr. Yaswant Singh Parmar University of Horticulture and Forestry, Solan
19. Birsa Agricultural University, Ranchi

Jammu and Kashmir

20. Sher-e-Kashmir University of Agricultural Science and Technology, Srinagar
21. Sher-e-Kashmir University of Agricultural Science and Technology, Jammu

Karnataka

22. University of Agricultural Sciences, Bangalore
23. Karnataka Veterinary, Animal and Fisheries Sciences University, Bidar
24. University of Agricultural Sciences, Raichur
25. University of Agricultural Sciences, Dharwad
26. University of Horticulture Science, Bagalkot
27. University of Agriculture and Horticulture Sciences, Shimoga

Kerala

28. Kerala Agricultural University, Thrissur
29. Kerala University of Fisheries and Ocean Studies, Panangad, Kochi
30. Kerala Veterinary and Animal Sciences University, Pookode, Kerala

Madhya Pradesh

31. Rajmata Vijayaraje Scindia Krishi Vishwa Vidyalaya, Gwalior
32. Nanaji Deshmukh Pashu Chikitsa Vishwa Vidyalaya, Jabalpur
33. Jawaharlal Nehru Krishi Viswa Vidyalaya, Jabalpur

Maharashtra

34. Dr. Balaesahib Sawant Kokan KrishiVidyapeeth, Dapoli
35. Maharastra Animal and Fisheries Sciences University, Nagpur
36. Vasantrao Naik Marathwada Krishi Vidyapeeth, Parbhani
37. Mahatma Phule Krishi Vidyapeeth, Rahuri
38. Dr. Punjabrao Deshmukh KrishiViswaVidyalaya, Akola

Odisha

39. Orissa University of Agricultural and Technology, Bhubaneswar

Punjab

40. Guru Angad Dev Veterinary and Animal Sciences University, Ludhiana
41. Punjab Agricultural University, Ludhiana

Rajasthan

42. Maharana Pratap University of Agriculture and Technology, Udaipur
43. Swami Keshwanand Rajasthan Agricultural University, Bikaner
44. Rajasthan University of Veterinary and Animal Sciences, Bikaner
45. SKN Agriculture University, Jobner
46. Agriculture University, Kota
47. Agriculture University, Jodhpur

Tamil Nadu

48. Tamil Nadu Agricultural University, Coimbatore
49. Tamil Nadu Veterinary and Animal Sciences University, Chennai
50. Tamil Nadu Fisheries University, Nagapattinam

Telangana

51. Sri Konda Laxman Telangana State Horticultural University, Hyderabad
52. Sri PV Narsimha Rao Telangana Veterinary University, Hyderabad
53. Professor Jayashankar Telangana State Agricultural University, Hyderabad

Uttrakhand

54. G.B. Pant University of Agriculture and Technology, Pantnagar
55. VCSG Uttarakhand University of Horticulture and Forestry, Bharsar

Uttar Pradesh

56. Chandra Shekhar Azad University of Agricultural and Technology, Kanpur
57. Narendra Deva University of Agriculture and Technology, Faizabad
58. Sardar Vallabhbhai Patel University of Agriculture and Technology, Meerut
59. U.P. Pt. Deen Dayal Upadhyaya Pashu Chikitsa Vigyan Vishwavidhyalaya Evem Go Anusandhan Sansthan, Mathura
60. Banda University of Agriculture and Technology, Banda
61. Sam Higginbottom University of Agriculture, Technology and Sciences, Allahabad

West Bengal

62. Bidhan Chandra Krishi Viswa Vidhyalaya, Mohanpur
63. West Bengal University of Animal and Fishery Sciences, Kolkata
64. Uttar Banga Krishi Viswavidhyalaya, Cooch Behar

Central Agricultural Universities

1. Central Agricultural University, Imphal, Manipur
2. Rani Laxmi Bai Central Agricultural University,Jhansi, Uttar Pradesh
3. Dr. Rajendra Prasad Central Agricultural University,Pusa (Samastipur)

AICRPS/Network Projects

1. AICRP on Nematodes, New Delhi
2. AICRP on Maize, New Delhi
3. AICRP Rice, Hyderabad
4. AICRP on Chickpea, Kanpur
5. AICRP on MULLARP, Kanpur
6. AICRP on Pigeon Pea, Kanpur
7. AICRP on Arid Legumes, Kanpur
8. AICRP on Wheat and Barley Improvement Project, Karnal
9. AICRP Sorghum, Hyderabad
10. AICRP on Pearl Millets, Jodhpur
11. AICRP on Small Millets, Bangalore
12. AICRP on Sugarcane, Lucknow
13. AICRP on Cotton, Coimbatore

14. AICRP on Groundnut, Junagarh
15. AICRP on Soybean, Indore
16. AICRP on Rapeseed and Mustard, Bharatpur
17. AICRP on Sunflower, Safflower, Castor, Hyderabad
18. AICRP on Linseed, Kanpur
19. AICRP on Sesame and Niger, Jabalpur
20. AICRP on IPM and Biocontrol, Bangalore
21. AICRP on Honey Bee Research and Training, Hisar
22. AICRP -NSP(Crops), Mau
23. AICRP on Forage Crops, Jhansi
24. AICRP on Fruits, Bangaluru
25. AICRP Arid Zone Fruits, Bikaner
26. AICRP Mushroom, Solan
27. AICRP Vegetables including NSP Vegetable, Varanasi
28. AICRP Potato, Shimla
29. AICRP Tuber Crops, Thiruvananthapuram
30. AICRP Palms, Kasaragod
31. AICRP Cashew, Puttur
32. AICRP Spices, Calicut
33. AICRP on Medicinal and Aromatic Plants including Betelvine, Anand
34. AICRP on Floriculture, New Delhi
35. AICRP in Micro Secondary and Pollutant Elements in Soils and Plants, Bhopal
36. AICRP on Soil Test with Crop Response, Bhopal
37. AICRP on Long Term Fertilizer Experiments, Bhopal
38. AICRP on Salt Affected Soils and Use of Saline Water in Agriculture, Karnal
39. AICRP on Water Management Research, Bhubaneshwar
40. AICRP on Ground Water Utilisation, Bhubaneshwar
41. AICRP Dryland Agriculture, Hyderabad
42. AICRP on Agrometeorology, Hyderabad including Network on Impact adaptation and Vulnerability of Indian Agri. to Climate Change
43. AICRP Integrated Farming System Research, Modipuram including Network Organic Farming
44. AICRP Weed Control, Jabalpur
45. AICRP on Agroforestry, Jhansi
46. AICRP on Farm Implements and Machinery, Bhopal

47. All India Coordinated Research Project on Ergonomics and Safety in Agriculture
48. AICRP on Energy in Agriculture and Agro Based Indus.,Bhopal
49. AICRP on Utilization of Animal Energy (UAE), Bhopal
50. AICRP on Plasticulture Engineering and Technologies, Ludhiana
51. AICRP on PHT, Ludhiana
52. AICRP on Goat Improvement, Mathura
53. AICRP- Improvement of Feed Sources and Nutrient Utilisation for raising animal production, Bangalore
54. AICRP on Cattle Research, Meerut
55. AICRP on Poultry, Hyderabad
56. AICRP-Pig, Izzatnagar
57. AICRP Foot and Mouth Disease, Mukteshwar
58. AICRP ADMAS, Bangalore
59. AICRP on Home Science, Bhubaneshwar

Network Projects - 19

1. All India Network Project on Pesticides Residues, New Delhi
2. All India Network Project on Underutilised Crops, New Delhi
3. All India Network Project on Tobacco, Rajahmundry
4. All India Network Project on Soil Arthropod Pests, Durgapura
5. Network on Agricultural Acarology, Bangalore
6. Network on Economic Ornithology, Hyderabad
7. All India Network Project on Rodent Control, Jodhpur
8. All India Network Project on Jute and Allied Fibres, Barrackpore
9. Network project on Improvement of Onion and Garlic, Pune
10. Network Bio-fertilizers, Bhopal
11. Network Project on Harvest and Postharvest and Value Addition to Natural Resins and Gums, Ranchi
12. Network project on Animal Genetic Resources, Karnal
13. Network Project on R&D Support for Process Upgradation of Indigenous Milk products for industrial application Karnal
14. Network Programme on Sheep Improvement, Avikanagar
15. Network Project on Buffaloes Improvement, Hisar
16. Network on Gastro Intestinal Parasitism, Izatnagar

17. Network on Haemorrhagic Septicaemia, Izatnagar
18. Network Programme Blue Tongue Disease, Izatnagar
19. Network Project on Conservation of Lac Insect Genetic Resources, Ranchi
20. Network Project on Agricultural Bioinformatics and Computational Biology, New Delhi

Other Projects - 10

1. Technology Mission on Cotton (CICR, Nagpur)
2. Technology Mission on Jute (CRIJAF, Barrackpore)
3. Continuation, Strengthening and Establishment of Krishi Vigyan Kendras
4. Strengthening and Development of Higher Agricultural Education in India, New Delhi
5. Central Agricultutral University, Imphal
6. Strengthening and Modernization of ICAR Headquarters
7. Intellectual Property Management and Transfer/Commercialisation of Agricultural Technology (Upscaling of existing component IPR HQ)
8. Indo US Knowledge Initiative
9. National Agricultural Innovative Project, New Delhi
10. National Fund for Basic and Strategic Research, New Delhi

Different National Boards

National Boards Name	*Establishment Year*	*Situated at*
Biodiversity board	2008	Chennai
Central silk board	1949	Banguluru
Coffee board	1942	Banguluru
Coconut development board	1981	Kochi
Fish and fish product board	2006	Hyderabad
Rubber board	1947	Kottayam, Kerala
Spice board	1986	Kochi
Tea board	1953	Kolkata
Tobacco board	1976	Guntur,A.P.

International Organizations of Crop Improvement

CGIAR	Consultative Group for International Agricultural Research	Washington (USA)
CIAT	Centro Internacional de Agricultura Tropical (International Centre for Tropical Agriculture)	Palmira, Columbia
CIFOR	Center for International Forestry Research	Jakarta, Indonesia
CIMMYT	Centro Internacional de Mejoramiento de Maizy Trigo (International Centre for Wheat and Maize Improvement)	Baton, Mexico
CIP	Centro Internacional de la Papa (International Potato Centre)	Lima, Peru
IBPGR	International Board for Plant Genetic Research in the Dry Areas Resources	Rome, Italy
ICARDA	International Center for Agricultural	Alleppo, Syria
ICGES	International Centre for Genetic Engineering and Biotechnology	Triesta, Italy and New Delhi, India
ICRAF	International Centre for Research in Agroforestry	Bagor, Indonesia
ICRISAT	International Crops Research Institute for the Semi-Arid Tropics	Hyderabad, India
IFPRI	International Food Policy Research Institute	Washington, USA
IITA	International Institute of Tropical Agriculture	Ibadan, Nigeria
IIMI	International Irrigation Management Institute	Colombo, Srilanka
ILRI	International Livestock Research Institute	Nairobi, Kenya
INSFFER	International Network on Soil Fertility and Fertilizer Evaluation on Rice	New Delhi, India
IPGRI	International Plant Genetic Resources	Rome, Italy Institute
ISNAR	International Service for National Agricultural Research	Netherlands
IRRI	International Rice Research Institute	Manila, Phillipines
IWMI	International Water Management Institute	Columbo, Sri Lanka
WFC	World Fish Centre	Bayan Lepas, Malaysia
WArDA	West African Rice Development Association	Monrovia, Liberia

Chapter 16

Current Scenario of Indian Agriculture

An Overview

- ☆ Agriculture has undergone significant developments since the time of the earliest cultivation. Involvement of revelation of plants and animals in agriculture was developed around 10,000 B.C.E., altering of communities of flora and fauna were already started by earlier people for their benefit by other means such as fire-stick farming (Bar-Yosef and Meadows, 1995).
- ☆ Each farming system has its pros and cons in past as well as in present scenario. Like nature, change is an avoidable rule in agricultural practices as well. From ancient to present time we have seen development of many such patterns to get better quality and quantity using available resources with new dimensions.

Agricultural Practices in India

Traditional Agriculture

- ☆ The evolution of traditional farming was over the foremost 10,000 years of agriculture. Traditional agriculture is basically sustainable and steady farming system that has been employed for a number of generations and is able to produce the material required by its producers. Through traditional farming a production of incredible variety of household crops and livestock, and systems of farming was made possible.
- ☆ Several conventional farmers in the developing world are still employing these farming methods that are in equilibrium with the nearby ecosystems,

steady, sustainable and highly organized. Conventional agriculture symbolizes the novel system of farming that developed through the interaction of societal and ecological systems. This system involves the rigorous use of local information and natural wealth supporting biological diversity by means of alternating practices. Conventional farmers centered on methods that preserve soil fertility, check the loss of topsoil, grip water in the soil and produce steady harvests.

Indian agricultural scenario has surely undergone many drastic changes and has achieved many milestones. The green revolution (1967-1978) transformed India from a food deficient stage to a surplus food market. In a span of 3 decades, India established itself as a net exporter of food grains. Interestingly, some developed countries, mainly Canada, which were facing a scarcity in agricultural labour, were so impressed by the results of India's Green Revolution that they showed interest in allowing farmers experienced in the methods of the Green Revolution to their own country. Many farmers from Punjab and Haryana states were then sent to Canada by GOI to settle there. That's why today one can see thousands of Punjabi-speaking citizens in Canada. Also in the Indian context, worth mentioning are the significant results achieved in the fields of dairying and oil seeds through our white and yellow revolutions respectively. As of now, in terms of agricultural output, India is ranked second in the world. India is also the largest producer in the world for milk, cashew nuts, coconuts, tea, ginger, turmeric and black pepper. India also boosts of the largest cattle population (193 million) in the whole world. Our country is also the second largest producer of wheat, rice, sugar, groundnut and inland fish. India is the third largest producer of tobacco. India is home to 10 per cent of the world fruit production with first rank in the production of fruits like banana and sapota. Presently, Indian Agriculture is witnessing a phase of diversification. During recent years, much awareness has been generated on shifting to high-yielding varieties (HYV) of crops from conventional crops. This has enabled a successful transition in Indian Agriculture from its stagnation to a growth path.

Types of Traditional Farming

- *Forest gardening:* Forest gardening is considered as one of the oldest agro-ecosystem, it is a plant- based food production system. Useful trees and plant species were identified, protected and improved and undesirable species were eliminated. Likewise superior species were selected and incorporated in family's garden.
- *Agro-Pastoralism:* In India agro-pastoralism incorporated threshing, planting crops in strips either of two or of six and storing up of grains in granaries.
- *Mixed farming:* Mixed farming was the basis of the Indus valley economy. In mixed farming arable farming is mixed with rising of livestock hence it is an agrarian system. Often the dung from the cattle is used as manure for increasing the production of cereal crops. Horses and cattle raised and used for haulage and bullocks to haul the cart and the plough.

- *Shifting cultivation:* It is also known as „slash and burn farming. It is a system in which forests are burnt for the release of nutrients which support for several years in the cultivation of annual as well as perennial crops. Then the plot is left as it is to grow again as a forest, mean while the farmer moves to another plot. Basically it is an ancient agricultural technique in which forests or woodlands are cleaned by burning of trees and plants so that the land can be used for farming. It is subsistence agriculture which typically uses little technology or other tools. It is a typical part of transhumance livestock herding and shifting cultivation agriculture.

Characteristics Features of Conventional Farming

- Traditional farming centre on threat diminution.
- Perennial vegetative cover of soils is attained through it.
- System diversity: Farm systems based on numerous cropping systems, and cropping systems are based on a combination of crops, and crops with varietal and other genetic variability upholds system diversity.
- Tropic complexity impending natural systems: Manifold interactions among plants, weeds, pathogens and insects occur in conventional farming because of which complexity occurs and remain maintained in natural systems.
- Traditional farming systems also promote genetic diversity: In India the *Mangifera indica* (mango), has been bred to produce 1,000 varieties, and some 100,000 varieties produced by one species of rice, *Oryza sativa*. India had more than 40 thousand varieties of rice, as of 1987, only 30 were commonly grown.

Demerits of Traditional Farming

- During the old days, agricultural technology was not well developed compared to now. People did not use any machinery to farm their land, machinery which are much more efficient replaced those animals used in the agriculture.
- Previously, farmers had to be depended on land suitability, climate factor and the availability of adequate water supply before planting a crop that suited those criteria.
- In the old days, farmers worked on their land to produce crops for their own domestic use because with the methods used in traditional agriculture were not very effective to produce huge amount of crops. The crop yield was really low according to the population in traditional agriculture.
- Pest management and disease management were not so effective in traditional agriculture.

- Due to environmental factors and undeveloped techniques overall production was low in traditional farming.

New Farming Systems

- Due to many challenges in traditional farming system, new techniques are used now days. In new techniques, organic farming, genetically modified crops, vertical farming, polytonal farming, greenhouse farming and multi-crop farming are main farming system. Two most important systems are described:

Organic Farming System

- Organic farming is a method of crop production that involves much more than choosing not to utilize pesticides, fertilizers, genetically modified organisms, antibiotics and growth hormones. It works in agreement with nature and involves using techniques to get good crop yields without harming the natural environment. Organic farming, evolved on the basic theoretical expositions of Rodale in the United States, Lady Balfour in England and Sir Albert Howard in India in the 1940s, has progressed to cover about 23 million hectares of land all over the world. In modern times, the general population is more aware about environmental benefit that's why organic farming is gradually getting popularity throughout the world.
- This system relies on such techniques like crop rotation, green manure, compost, and biological pest control. This system focused on the appropriate application of soil organic matter to augment soil fertility. Animal dung, crop residue, green manure, food processing wastes and biofertilizers are some resources which are usually used for enhancing soil fertility in this system.
- There are many types of organic farming system which are mentioned here:
    - Biodynamic farming was developed by Rudolf Steiner. This method emphasizes the use of manuresand composts and excludes the use of artificial chemicals on soil and plants. It treats soil fertility, plant growth, and livestock care as ecologically interrelated tasks.
    - Do nothing or Natural farming is established by Masanobu Fukuoka, Japanese farmer. This farming is a closed system, in which system demands no inputs and imitates nature. This farming have refined into five ideologies that are no use tillage, fertilizer, pesticides, weeding and pruning.
    - Bio-intensive farming system is based on closed system which main aims maximum crop yield from minimum area of land. This system is focused on simultaneously improving and maintaining the fertility of the soil.

- Other systems of this organic farming are holistic management, permaculture, SRI and no-till farming.

Benefits of Organic Farming System

- ***Crop yield and plant diversity:*** Many reports have shown that higher yield in organic agriculture system observed than comparable traditional agriculture. Although, many studies reported that in organic transition effect period (first 1-4 year of organic agriculture in a land) crop yield decline, followed by a yield increment with well developed biological activity of soil. Now days, this system seems to be beneficial for biodiversity. Organic methods of this farming such as rotating crops to build soil fertility and naturally raising animals helps to promote biodiversity.
- ***Organic farming and environmental earnings***: Many reports have indicated that significant environmental amelioration through conversion from conventional farming to organic farming is awesome. Organic farming is a system in which chemical and pesticides are avoided for crop production. So, organic farming discourages environmental exposure to these harmful inputs and fight for environment protection.
- **Insect and ailment control in organic farming:** In organic farming system, prevention from pest and disease is done by means of naturally resistant crops against disease and pest or by selecting suitable sowing times that avoid pest and disease and subsequent pest epidemics. Other methods which generally applied to control pest and disease attack are; encouraging natural biological agents for control of diseases, insects; rotating crops; by using semi chemical for trapping pests.
- **Nutrient management and safety of organic foods:** In recent time, organic farming system is getting value and popularity due to its positive results in nutrient management and safety of food produced. Crop rotation and use of varieties of crops are suitable method in this system because of potential to balance the demand of nitrogen for crops.

Genetically Modified Crops System

- Genetically modified crops (GM crops) are crops of which, the DNA has been modified using various genetic engineering techniques. The aim of making GM crops is to set up a new characteristic to the plant which does not occur in nature in particular plant species *e.g.* in food crops various pest resistant, abiotic stress resistant, spoilage resistant, chemical resistant genes have been incorporated along with those genes which have improved the nutrient profile of the crop. In India this system is not very popular because of serious socio-economical reasons but the system is getting attention at the level of research for various factors.

Making of Genetically Modified Crops

GM crops are developed in a laboratory conditions by altering their genetic configuration. This is generally done by adding up one or more genes to a genome of selected plants using various gene manipulating methods. However, most of the GM crops by and large are developed using biolistic method (particle gun) or *Agrobacterium tumefaciens* mediated transformation method.

The best known example of making of GM crops is the use of *Bt* (*Bacillus thuringiensis*) genes in the crops like corn and cotton. It is a naturally occurring bacterium that produces crystal (cry) proteins which are fatal to insect larvae. *Bt* crystal protein genes have been transferred into crops, facilitating the crops to produce its own pesticides against insects like corn borer. Tobacco and *Arabidopsis thaliana* are the most accepted GM plants due to well setteled and trusted methods of transformation, effortless propagation and well studied genomes.

Advantages of Genetically Modified Crops

GM crops grown or under trial development, have been tailored with traits according to benefit of farmers, customers, and industry. In recent times, few researches have been targeted the development of crops which will be important for developing nations such as insect resistant brinjal for India and insect-resistant cowpea for Africa.

Some major advantages are as follows:

- ☆ Stress resistance
- ☆ Pathogen resistance
- ☆ Improved nutrition
- ☆ Herbicide resistance
- ☆ Production of biofuels
- ☆ Role in Bioremediation

Organic Farming System Verses Genetically Modified Crops

Organic farming system and GM crops both have their characteristic value on the basis of their benefits for human being. But many environment activists are against the use of GM crops.

In India, many objections are coming in the light by which organic farming is seems more beneficial than GM crops in following aspects:

- ☆ Unintended harm to non-targeted organisms is major disadvantage of genetically modified crops: For example in BT crops, produced *Bt* toxins, kill many species which are not harmful for crops. This study was re-scrutinized by the USDA, the U.S. Environmental Protection Agency (EPA) by which concluded that this fact may be imperfect, while in case of organic farming system, farmers never face such problem.

- Genetic contamination -Possibilities of desired gene transfer into non-targeted species are much high in GM crops: For example if GM crop of herbicide tolerance and weed will cross breed, ensuing in the transfer of the genes for herbicide resistance from source crops into weeds. These "superweeds" would be then become herbicide tolerant which would be harmful for crops. On the other hand, this kind of problems is not faced in organic farming system.
- Human health issue is very controversial issue in the regarding of GM crops: Allergenicity is one of the most negative aspects of GM crop uses. Many children in US and Europe have affected with allergy to transgenic peanut and other food. That s why proposal of making transgenic soybean through Brazil nuts gene was dump. Other unknown effect also observed on health like effect of GM potatoes on digestive tract on rats but these studies are not fully confirmed yet. Moreover, a gene of snowdrop flower, lectin introduced into potatoes is known as toxic for humans so these potatoes were never used for humans. All these kind of issues are not faced in organic based farming based system because there are many kind of natural source used for nutrient management of crops.
- According to economics concern, bringing of GM food to market is a very extensive and expensive process. Many researches are remains to solve unidentified risk of GM foods with human health which are time taking and costly. So GM crops are not economically favorable. On the other hand, organic farming is seems to be favourable to economic point of any country because of less financial input.

Agriculture Schemes in India

- *National Food Security Mission (NFSM):* To increase the production of rice, wheat and pulses. The mission is being continued during 12th plan with new target of additional production of 25 million tonnes of foodgrains comprising 10 million tonnes of rice, 8 million tonnes of wheat and 4 million tonnes of pulses and 3 million tonnes of coarse cereals.
- *National Food Security Mission-Commercial Crops:* For crop development programme on cotton, jute and sugarcane for enhancing productivity.
- *Mission for Integrated Development of Horticulture (MIDH):* It covers wide horticulture base, which includes fruits, vegetables, tuber crops, mushrooms, spices and aromatic plants flowers and foliage and plantation crops like coconut, arecanut, cashew nut, cocoa and bamboo.
- *National Mission on Oilseeds and Oil Palm:* Envisages increase in production of vegetable oils sourced from oilseeds, oil palm and tree borne oilseeds.
- *National Mission for Sustainable Agriculture:* Aims at making agriculture more productive, sustainable and remunerative and climate resilient by promoting location specific integrated/composite farming systems;

soil and moisture conservation measures; comprehensive soil health management; efficient water management practices and mainstreaming rainfed technologies.

- *National Mission on Agricultural Extension and Technology:* Its aim is to restructure and strengthen agricultural extension to enable delivery of appropriate technology and improved agronomic practices to the farmers consists.
- *Initiative for increasing flow of credit:* In order to ensure that all eligible farmers are provided with hassle free and timely credit for their agricultural operation, **Kisan Credit Card** (KCC) Scheme was introduced in 1998-99. The main objectives of the scheme are to meet the short term credit requirements for cultivation of crops, post harvest expenses, produce marketing loan, consumption requirements of farmer household, working capital for maintenance of farm assets and activities allied to agriculture like dairy animals, inland fishery, *etc.*, investment credit requirement for agriculture and allied activities like pump sets, sprayers, dairy animals, *etc.*
- *Pradhan Mantri Fasal Bima Yojana:* Under the new scheme, the farmers' premium has been kept at a maximum of 2 per cent for foodgrains and oilseeds, and up to 5 per cent for horticulture and cotton crops. There is no upper limit on Government subsidy. Even if balance premium is 90 per cent, it will be borne by the Government. Earlier, there was a provision of capping the premium rate which resulted in low claims being paid to farmers. This capping was done to limit Government outgo on the premium subsidy. This capping has now been removed and farmers will get claim against full sum insured without any reduction. Importantly for the beneficiaries, crop losses which are covered under the scheme include Yield Losses as well as post-harvest losses, where coverage will be available up to a maximum period of 14 days from harvesting for those crops. The use of technology will be encouraged to a great extent resulting in operational efficiency. Smart phones will be used to capture and upload data of crop cutting to reduce the delays in claim payment to farmers. Remote sensing will be used to reduce the number of crop cutting experiments.
- *Mera Gaon, Mera Gaurav:* This scheme is being launched involving agricultural experts of agricultural universities and ICAR institutes for effective and deeper reach of scientific farming to the villages. A group of experts will be associated with one particular village to create awareness and adoption of new technologies including farm investment, loans, availability of inputs and marketing. All the scientists from ICAR and agricultural universities will participate in this initiative.
- *Krishi Dak:* IARI initiated this novel scheme in 20 districts in which postmen supplied seeds of improved varieties of crops to the farmers in

far-flung areas. Owing to its success and popularity, this scheme is being extended in 100 districts of 14 states with the association of Krishi Vigyan Kendras. This will provide improved seed to farmers at their doorstep.

- *Soil Health Card:* Soil Health cards are necessary to ensure that only requisite nutrients are applied in the soil in a balanced manner to enhance productivity of specific crops in a sustainable manner. Values on soil parameters such as pH, EC, N, P, K, S, Zn, Fe, Mn, Cu and B. Recommendation on appropriate dosage of fertilizer application based on test values and requirement of crop, use of organic manures and soil amendments to acidic/alkaline/sodic soils.
- *Paramparagat Krishi Vikas Yojna (PKVY):* Aim of the project is to maximize the utilization of natural resources through eco-friendly cultivation. Organic farming is a method of farming system which primarily aimed at cultivating the land and raising crops in such a way, as to keep the soil alive and in good health by use of organic wastes (crop, animal and farm wastes, aquatic wastes) and other biological materials along with beneficial microbes (bio-fertilizers) to release nutrients to crops for increased sustainable production in an eco friendly pollution free environment.
- *Promotion of National Market through Agri Tech Infrastructure Fund (ATIF):* Central Sector Scheme for Promotion of National Agricultural Market through Agri-Tech Infrastructure Fund (ATIF) for Rs.200 crores to be implemented during 2014-15 to 2016-17. The Scheme envisages initiation of e-marketing platform at the national level and will support creation of infrastructure to enable e-marketing in 642 regulated markets across the country. For creation of a National Market, a common platform across all States is necessary. It is, therefore, proposed that a service provider be engaged centrally who would build, operate and maintain the e-platform on PPP (Build, Own, Operate, Transfer – BOOT) model. This platform would be customized/configured to address the variations in different states. As an initiative of deregulation, States have been advised by the Government of India to bring fruits and vegetables out of the ambit of APMC Act. In pursuance of this advisory, 12 States have, so far, either de-regulated the marketing of fruits and vegetables or have exempted from levying of market fee.
- *MUDRA Bank*: The Finance Minister has proposed to create a Micro Units Development Refinance Agency (MUDRA) Bank, with a corpus of Rs. 20,000 crore, and credit guarantee corpus of 3,000 crore, which will refinance Micro-Finance Institutions through a Pradhan Mantri Mudra Yojana. Priority will be given to SC/ST enterprises in lending. MUDRA Bank will operate through regional level financing institutions who in turn will connect with last mile lenders such as MFIs, Small Banks, Primary Credit Cooperative Societies, Self Help Groups (SHGs), NBFC (other than

MFI) and other lending institutions. MUDRA Bank will refinance Micro-Finance Institutions through a Pradhan Mantri Mudra Yojana (PMMY). In lending, priority will be given to SC/ST enterprises. These measures will greatly increase the confidence of young, educated or skilled workers who would not be able to aspire to become first generation entrepreneurs; existing small businesses, too will be able to expand their activities. Since the MUDRA Bank will be set up through an enactment of law and it will take some time.

- *Krishi Kalyan Cess:* Union Budget for 2016-17 (April-March) introduced a new cess on services named Krishi Kalyan Cess at the rate of 0.5 per cent. The effective rate of the Krishi Kalyan Cess, however, will be lower than 0.5 per cent as the government will provide input tax credit for the cess, as against no input tax credit for Swachh Bharat Cess. Proceeds of cess would be exclusively used for financing initiatives relating to improvement of agriculture and welfare of farmers.
- *Direct Benefit Transfer (DBT) for Fertilizer Sector:* Government also announced to introduce Direct Benefit Transfer of fertilizer subsidy to farmers on pilot basis in few districts of the country.

Courses and Future Prospects

Introduction to Indian Agriculture Courses

- Well, as we all know India is agriculture based country, it is essential to pursue agricultural education to boost the country's economy. Apart from this, to reach the demand in food sector, it needs more research oriented programs in agriculture sector.
- Indian population is growing very fast, but agriculture related products are not reaching the demand, this make us to think and opt out for agriculture related courses which will have great demand in coming days.
- Usually, once agriculture degree or course is completed, you can find a job in Food Science and Technology, Agriculture Universities (As a research scholar or staff), Dairy Farming, Agricultural Marketing, State and Central Horticulture department, Agriculture/Forest Department and other such related fields.
- We are presenting here about agriculture courses details and eligibility to help you in making the right decision in pursuing in your agriculture career which can also help Indian agriculture sector and economy.
- There are many universities and colleges/institutes in India are offering different agriculture courses at different levels.

Table 16.1: Indian Agriculture Courses

Certificate Course in Agriculture Science	This would be a 1 year course	Any student can join after passing the 10th class exam.
Diploma in Agriculture	This would be a 2 year course	Any student can join after passing the 10th class exam.
B.Sc. in Agriculture.	This would be a 4 year Graduate Degree program	Any student who has passed 10+2 (Science Stream) with 55 per cent of marks is eligible to appear for entrance. You can select one of these: Agricultural Sciences, Agricultural Economics, Agricultural Engineering, Agricultural Entomology, Agricultural Microbiology, Agricultural Statistics, and Agronomy.
B.Sc. (Hons.) in Agriculture	This would be a 3 year graduate degree program	Any student can join after passing the 10+2 (Science Stream) exam.
BBA Agriculture	This would be a 4 year Graduate degree program	Any student can join after passing the 10+2 exam.
M.Sc. in Agriculture	This would be a 2 year Post- graduation Degree course	To eligible, must be graduated with specialization in Agriculture, Agricultural Biotechnology, Agricultural Business Management, Forestry, Horticulture or any other related discipline.
MBA (International Agribusiness)	This would be a 2 year Graduate Degree program	Students after completing graduation in any discipline.
M.Tech. (Dairy Technology)	This is would be a 2 year Post- graduate degree program	Students must be graduated in an Agricultural discipline.
M.Phil. (Horticulture)	This would be a 2 year Pre- doctorate Degree level program (course)	Must have Master's degree in Horticulture.
Ph.D. in Agriculture	This would be a 3 year Doctorate level program	Must be Post- graduated in an Agricultural discipline.

Indian Agriculture

- ✰ Agriculture in India has been a profession for thousands of years
- ✰ Presently country holds the second position in agricultural production all over the world
- ✰ Indian agriculture has made rapid strides since independence:
  - ❀ From food shortages and import to self sufficiency and exports
  - ❀ From subsistence farming to intensive and technology led cultivation

- Today India is the front ranking producer of many crops in the world
- Ushered in through the green white, blue and yellow revolutions.

- Key Drivers of Agriculture
  - Technology (farming and crop)
  - Government Policy (credit, crop specific programmes)
  - Cropping pattern (which depends on profitability, awareness *etc.*)
  - Environment factors (water availability, soil degradation, climate change *etc.*)
  - Market forces (market openness, prices, transparency, integration with downstream sectors)
  - Global factors (supply-demand, trade norms, and restrictions *etc.*)

Table 16.2: Indian Agriculture in Global Ranking

Total Area	7th
Irrigated Area	1st
Population	2nd
Economically Active Population	2nd
Total Cereals	3rd
Wheat	2nd
Rice	2nd
Coarse grains	4th
Total Pulses	1st
Oilseeds	2nd
Fruits and Vegetables	2nd
Milk	1st
Livestock (cattle, buffaloes)	1st
Implements (Tractors)	3rd

Major Indian Agricultural Concerns

- Small and fragmented land holdings
- Imbalanced use of Fertilizer and Pesticides
- Shortage of good quality seeds especially for Small and Marginal Farmers
- Problem of Irrigation – wastage of Water on the one hand and scarcity of Water on the other
- Soil Erosion
- Lack of PHM and Marketing Facilities – Storage, Transport and Cold Chain
- Scarcity of Capital

Lack of new technologies post Green Revolution have worn off which is bcoming worrisome (BT cotton being the exception). Very few promising seeds have been commercialised and most other innovations are still languishing due to poor extension or lack of investment. Corrective action regarding fertilizer subsidy, inadequate seed production, market rigidities and other market-distortion policies is desperately needed. Achievement of irrigation potential given the large backlog of previous projects need to be completed. Monsoon dependence and erratic growth will continue to plague the sector. Fundamental degradation in environmental parameters and lack of good practices may impact future food security. A consistent movement of labour away from agricultural occupations may lead to rising agricultural wages. This should prompt greater mechanization and productivity enhancement measures.

Future Prospects

- India's experience with contract farming has been universally positive in raising profitability – this is being helped by the emergence of organised food retail
- However, there are issues around reneging of contracts and political opposition that need to be addressed
- Reform in the land policy for accelerated agricultural growth and poverty reduction
- A paradigm shift from subsistence farming to market oriented commercial agriculture
- Improve viability of small and marginal farms
- Enable small farms to benefit from agricultural diversification and value addition

Opportunities

There are several opportunities which can help revival of agriculture like:

- Contract Farming
- Reforming Land Policy
- New Crop Technologies – Seed Fertilizer Crop Protection
- PHM, Retail and Distribution.

Glossary

Abadi: Carries the general sense of populated and cultivated country, population and cultivation necessarily going together. Used to describe a condition, it is best rendered as "prosperity": when applied to a process, it denotes "development.

Abiota: The nonliving component of an ecosystem, including the soil, water, and air.

Aborigines: The original or native inhabitants of a country or region.

Agricultural revolution: A shift that took place 10,000 to 12,000 years ago and was characterized by the movement of human activity from hunting and gathering to agriculture.

Alternative crops: Non-traditional crops that can be grown in an area to diversify rotations and increase income.

Amin: An official designation. Under Sher Shah, probably one of the two chief officials in a pargana (but see under Amir). Under Akbar, an official on the staff of a Viceroy, whose precise duties are not explained. In 170., a revenue assessor under the provincial *Diwan*. May also, apparently. be used in a wider sense to denote an officer's "deputy" or "assistant."

Ancient forests: Forests that have never been cut and typically consist of trees 250 years of age and older.

Arable land: Land that can be cultivated.

Arid: A condition in which less than 10 inches of rain falls each year and the level of evaporation is greater than the level of precipitation.

Batai: Sharing produce by Division.

Bigha: The ordinary unit of area; its size varied within very wide limits, both by place and by period.

Biswa: One-twentieth of a bigha.

Black Death: The name for the plague that swept Europe and Asia during the Middle Ages.

Chaudhri: The headman of a pargana.

Chauth: The claim, nominally one-fourth of the revenue, made by the Marathas on country which they overran, but did not administer.

Community: A group interacting at a specific time and place, as well as sharing a similar cultural background.

Compost: Decayed organic and animal matter used as fertilizer.

Cover crops: Plants used to hold the soil during the fallow season.

Crop rotation: An agricultural method in which two or more crops are rotated from year to year to reduce nutrient depletion of the soil and reliance on pesticides.

Daftar: A record. *Daftarkhana*, record office.

Dastur: Has various general senses, "custom," "permission," "a Minister." Under Akbar and later, a schedule of assessment-rates stated in money; an abbreviation of *dastur-ul 'amal.*

Deh: A village in the Indian sense, which is nearly that of "civil parish," that is, a small area recognised as an administrative unit, not necessarily inhabited: synonyms, Mauza, Qariyat.

Desert: A biome with limited precipitation (typically less than 10 inches per year) but different temperature ranges. Tropical deserts (Sahara) are hot year round, temperate deserts (Mojave) are hot in the summer and cool in the winter, and cold deserts (Gobi) are hot or warm in the summer and cold in the winter.

Detritivore: An organism that feeds on dead organic matter.

Dfwan, Diwani: Discussed in Introduction. In 13-140., *Diwan* meant a Ministry. In 16c., (I) the Revenue Minister, (2) a nobleman's steward. In 17c., (1) a high official in the Revenue Ministry, (2) the provincial Revenue Officer. *Diwani* in 16c. meant the Revenue Ministry; in 17c, and later the revenue and financial administration as a whole; in 19c., the Civil Courts.

Doab: A region lying between two rivers, especially that between the Ganges and the Jumna.

Erosion: A process by which rock particles and soil are deposited in a new location through water or wind action.

Famine: Malnutrition and starvation resulting from a shortage of food.

Fauna: The animal life of an area.

Forest: A biome with enough precipitation to support various tree species.

Fossil fuels: Mineral fuels that occur in rock formations. They include coal, oil, and natural gas and provide a majority of the energy used in the world.

Gram: Anglicised from Portuguese *grao*: a pulse *(Cicer arietinum).*

Grassland: A biome with a moderate level of precipitation and vegetation dominated by grasses.

Gumashta: An assistant or subordinate. In the *Ain*, applied to subordinates employed by the Collector in Reserved land.

Hage (Haqq): In addition to the general senses-right, justice, truth, etc.–denoted in 13-14c., the perquisites allowed to Chiefs, usually in the form of land free from assessment. Haqq-i shirb, a term of Islamic law, denoting the right accruing to a person who provided water for irrigation.

Ijara: A farm of revenue. The farmer is usually *Ijaradar*; also *Mustajir*.

Inam: A reward. Applied specially to gifts made by the King, whether in the form of a sum of money, or a stipend paid in cash, or a Grant of revenue.

Integrated pest management: A management effort combining biological and chemical controls to reduce reliance on synthetic pesticides.

Intensive agriculture: A system to maximize output of land through use of chemicals and machinery.

Irrigation: Artificial watering of crops.

Jagir: An Assignment or revenue.

Jarib: A land measure; also, the measuring instrument. In 16c., used to denote assessment by Measurement, as synonym of *Paimaish*.

Jowar: A millet. *(Andropogon sorghum.)*

Karavaniyan: Used by Barni to denote the itinerant grain merchants, usually called *Banjaras*.

Kharif: The rains season, and the crops grown in it.

Kroh: A measure of distance, about 1 miles.

Kror: Ten millions (100 lakhs).

Madad-I Maash: A Grant of land for subsistence. Mahal (Mahal). Under Akbar, a revenue-subdivision, corre sponding usually, but not invariably, with pargana; and occasionally applied also to a head of miscellaneous revenue. The modern form, *mahal*, does not appear before 18c.

Malik. Carries the general idea of sovereignty or dominion In Islamic law, applied to an occupant of land, and used in one of Aurangzeb's *farmans* to denote a peasant. *Malikana*, in the British period, denotes an allowance made to a landholder, or claimant, excluded from possession.

Masahat: Measurement, Survey. In 14c., denoted the process of assessment by Measurement, which in later times was called *Jarib*, or *Paimaish*.

Masha: An Indian weight, equal to 15 grains.

Mauza: In 13c., used generally in a wide sense as a place or locality; later, denotes a village in the Indian sense); synonym of Deh.

Moth: A pulse *(Phaseolus aconitifolius).*

Nasaq: The general sense is "order" Under Akbar, applied to a particular or "administration." form of revenue-administration, which I identify with In 16c., denoted the Group-assessment, though it may cover also Farming.

Natural resources: Substances and processes used by people that they cannot create.

Old-growth forests: Forests consisting of trees 250 years or older in age.

Paimaish: Measurement process of assessment by Measurement, as a synonym for *Jarib*.

Pargana: The Indian name for an aggregate of villages. Came into official Moslem use in 14c., partially superseding

Patwari: The village-accountant, a Hindi term adopted from the outset in Moslem administration.

Poverty: A low standard of living in terms of other people (relative poverty) or in terms of basic needs (absolute poverty).

Qabuliyat: Written undertaking given for the payment of revenue; the counterpart of a *Patta*.

Qasba: *Patta* (Lease), the document given to a revenue payer, indicating the sum which he had to pay.

Rabi: In India, the winter; the crops grown in winter and harvested in spring.

Rangeland: An area that provides vegetation for grazing animals.

Sadr: In the Mogul period, the designation of a high officer whose duties included the supervision of Grants.

Sarkar: In the chronicles usually means a treasury, whether belonging to the king or to a noble. Under Sher Shah, denoted an administrative district, *i.e.* an aggregate of Parganas: under Akbar, a revenue-district.

Savanna: A biome similar to grasslands, but receiving more precipitation and containing more trees.

Suba: In the Mogul period, a province of the Empire.

Suyurghal: In the Mogul period, allowances granted by the Emperor, whether paid in cash, or by Grants of land.

Topsoil: The top layer of the soil that contains large amounts of organic matter. It often is viewed as a transitional region between the living and nonliving segments of the larger biosphere.

Trace gases: Gases that occur in only small amounts.

Traditional agriculture: Farming based on practices such as crop rotation, use of animal manures instead of chemical fertilizers, and use of animal power.

Ushr: The tithe levied under Islamic law. Ushri denotes country liable to tithe, as opposed to *kharaji.*

Wiran: Deserted. Applied to a village which had been abandoned and was uncultivated.

Zabt: In Akbar's time, the system of assessment by Measurement as then practised. The adjective *zabti* was used to denote an area where the system was in force. In later times *zabti* denoted a revenue rate, or rent-rate, levied on the area sown, and varying with the crop.

Zamindar: Lit. "land-holder." The word does not necessarily imply any particular claim or title, and in 18c. was used in Bengal to denote any sort of holder (vide Ch. VII, sec. 2). In the literature of North India, from 14c. onwards, it meant what I have called a Chief, that is, a landholder with title or claim antecedent to Moslem rule, commonly a Raja, Rao, or some other Hindu King, or ex-King, who had become tributary to the Moslem State. It is occasionally applied also to rulers who had not become tributary.

References

Acquaah, G. (2002). Agricultural Production Systems: "Principles of Crop Production, Theories, Techniques and Technology". Prentice Hall, Upper Saddle River, NJ. pp. 283–317.

Akbar, Razia. (Tr.) 2000. *Nuskha Dar Fanni-Falahat* (The Art of Agriculture). Agri-History Bulletin No.3. Asian Agri-History Foundation, Secunderabad 500 009, India. 136 pp.

Alam, A. (2014). Soil degradation: A challenge to Sustainable Agriculture. *International Journal of Scientific Research in Agricultural Sciences*, 1(4): 50-55.

Altieri, M.A. (1987). Agroecology, The Scientific Basis of Alternative Agriculture. Westview Press, Boulder, Colorado.

Azcon Aguilar C. and Barea J.M. (1996). Arbuscular mycorrhizas and biological control of soil-borne pathogens – An overview of the mechanism involved. *Mycorrhiza*, 6: 457–464.

Baber, Z. (1996). The Science of Empire: Scientific Knowledge, Civilisation, and Colonial Rule in India. State University of New York Press, Albany.

Balambal, V. 1993. Agriculture in the Sangam age. In: *Agriculture in Ancient India. Itihas Patrika Prakashan*, Thane 400 602, India. pp. 26-39.

Banjara M., Zhu L., Shen G, Payton P and Zhang H (2012). Expression of an Arabidopsis sodium/proton antiporter gene (AtNHX1) in peanut to improve salt tolerance. *Plant Biotechnology Report*, 6 :59–67

Bansil, P.C. (2017). Economic Problems of Indian Agriculture. Daya Publishing House, A Division of Astral International Pvt Ltd, New Delhi, India.

Bansil, P.C. (2018). Floriculture: Cultivation, Processing and Marketing. Daya Publishing House, A Division of Astral International Pvt Ltd, New Delhi, India.

Bansil, P.C. (2018). History and Development of Indian Agriculture. Daya Publishing House, A Division of Astral International Pvt Ltd, New Delhi, India.

Behera K.K., Alam A., Vats S., Sharma V. and Sharma H.P. (2012). Organic Farming History and Techniques. In: E. Lichtfouse (ed.), Agroecology and Strategies for Climate Change, Sustainable Agriculture Reviews 8, DOI 10.1007/978-94-007-1905-7_12, Springer Science+Business Media B.V. 2012.

Bessey, E.A. 1950. *Morphology and Taxonomy of Fungi*. Constable and Company Ltd., London UK. 791 pp.

Bhagwagar, P.R. and Patel, M.K. 1999. Review of work on fungicides in India up to 1940. *Asian Agri-History* 3:67-82.

Bhat, M.R.1981. Varahamihira's *Brhat Samhita*. Part 1. Motilal Banarsidass, Delhi 110 007, India. pp. 527-535. (Reprint 1992.)

Bowman A.K. and Rogan E. (1999). Agriculture in Egypt: From Pharaonic to Modern Times. Oxford University Press, India.

Boyle, R. (2011). How to Genetically Modify a Seed, Step by Step. Popular Science

BSIP. 2000. Birbal Sahni *Institute of Palaeobotany Annual Report*. 1999-2000. BSIP, Lucknow, India. 73 pp.

Burkhardt P. (1996). Genetic engineering towards carotene biosynthesis in endosperm. Swiss Federal Institute of Technology.

Burton S. (1998). A History of India. Blackwell Publishing, United Kingdom, CSIRO 2006 press release on SeedQuest website.

Chand, Devi. 1997. *The Atharvaveda*. (In English.) Munshiram Manoharlal Publishers Pvt. Ltd., New Delhi, India. 939 p.

Deshmukh, I. (1986). Ecology and Tropical Biology. Blackwell Scientific Publications, Oxford.

Dev, S.M. (2006). Agricultural Labor and Wages since 1950. Encyclopedia of India (vol. 1) edited by Stanley Wolpert, pp. 17–20, Thomson Gale, ISBN 0-684-31350-2.

Enserink, M. (1999). The Lancet Scolded Over Pusztai Paper. *Science Magazine*, 286 (5440): 656.

Firminger, T.A.C.1864. *Manual of Gardening for Bengal and Upper India*. RC. Lepage and Co., London, UK and Calcutta, India. 558 pp.

Fukuoka, M. (1975).The One-Straw Revolution: An Introduction to Natural Farming. 1978 representation. In: Pearce C, Kurosawa T, and Korn L(tr. ed.) Rodale Press, New York.

Green, T. (1987). Organic Farming in India. *Alternatives* 15(1): 4-13.

Grigg, D.B. (7 November 1974). The Agricultural Systems of the World: An Evolutionary Approach. Cambridge University Press. ISBN 978-0-521-09843-4. Retrieved 2 May 2013.

Gulati, A. (2006). Agricultural Growth and Diversification since 1991. Encyclopaedia of India (vol. 1) edited by Stanley Wolpert, pp. 14–17, Thomson Gale, ISBN 0-684-31350-2.

Gupte, R.S., and Raje, R.A. 1896. *Krishikarmavidya*. (In Marathi.) T.K. Gajjar, Bombay New Press, Bombay (Mumbai), India. 740pp.

Javalia, B.M.1999. Recommendations related to arbori-horticulture in 19th century Rajasthan: *Extension of Vrikshayugnyana in Agnipurana. Asian Agri-History* 3:261-269.

Jha, V. 1999. *Amarkosha*. (In Hindi.) Volumes I, II, and III. Motilal Banarsidass, Delhi 110 007, India.

Kler D.S., Uppal R.S., Kumar A. and Kaur G. (2002). Imbalance and maintenance of eco-system– A review. Environment and Ecology, 20: 20–48.

Kler D.S., Kumar A., Chhinna G.S., Kaur R. and Uppal R.S. (2001). Essentials of organic farming– A review. Environment and Ecology, 19: 776–798.

Kler D.S., Singh S. and Walia SS (26-30 November 2002). Studies on organic versus chemical farming. Extended summaries vol. 1, 2nd International Agronomy Congress, New Delhi, pp. 39–40.

Koornneef M. and Meinke D. (2010). The development of Arabidopsis as a model plant. *The Plant journal*, 61 (6): 909–921.

Krishnamurthy, Radha. 1993. Agriculture in ancient India and its social aspects. In: *Agriculture in Ancient India*. Itihas Patrika Prakashan, Thane 400 602, India. pp. 40-55.

Liebhardt W.C., Andrews R.W., Culik M.N., Harwood R.R., Janke R.R., Radke J.K. and Rieger-Schwartz S.L. (1989). Crop production during conversion from conventional to low-input methods. *Agronomy Journal*, 81: 150–159.

Lochhead, C. (30 April 2012). Genetically modified crops' results raise concern. The San Francisco Chronicle.

Lynda, S. (2000). Southernisation. In: Adas M (ed). Agricultural and Pastoral Societies in Ancient and Classical History. Temple University Press, United State. pp- 308-324.

MacKenzie D. (18 June 1994). Transgenic tobacco is European first. New Scientist.

Majumdar, G.P. 1935. *Upavana-Vinoda* (A Sanskrit Treatise on Arbori-Horticulture). Indian Research Institute, Calcutta, India. 128 pp.

Meagher R.B. (2000). Phytoremediation of toxic elemental and organic pollutants. *Current Opinion in Plant Biology*. 3 (2): 153–162.

Mehra, K.L. (1997). Biodiversity and subsistence changes in India: The Neolithic and Chalcolothic Age. *Asian Agri-History* 1: 105-126.

Mirret E.P. (2001). Trees on Organic Farms. North Carolina State University.1-37.

Muramoto J. (2008). Comparison of nitrate content in leafy vegetables from organic and conventional farms in California. Organic Farming Research Foundation Newsletter, 8, www.ofrf.org.

Myrdal J. and Morell M. (2011). The Agrarian History of Sweden: From 4000 BC to AD 2000. Nordic Academic Press. p. 265. ISBN 9185509566. Retrieved 25 June 2013.

Naqvi S., Zhu C., Farre G., Ramessar K., Bassie L., Breitenbach J., Conesa D.P., Gaspar Ros G., Sandmann G., Capell T. and Christou P. (2009). Transgenic multivitamin corn through biofortification of endosperm with three vitamins representing three distinct metabolic pathways. 1-6. doi: 10.1073/pnas.0901412106.

National Academy of Sciences (2001). Transgenic Plants and World Agriculture. National Academy Press, Washington.

Neera P., Katano M. and Hasegawa T. (1999). Comparison of rice yield after various years of cultivation by natural farming. *Plant Production Science*, 2: 58–64.

Nene, Y.L. (1998b). Jahangir: A naturalist - III. Description of flora. *Asian Agri-History* 2:227-242.

Nene, Y.L. (1999). Seed health in ancient and medieval history and its relevance to presentday agriculture. *Asian Agri -History* 3: 157-184.

Nene, Y.L. and Sadhale, Nalini. (1997). Agriculture and biology in *Rigveda*. *Asian Agri-History* 1:177-190.

Nene, Y.L. (1998a). Jahangir: A naturalist - I. Environment and general observations. *Asian Agri-History* 2:25-35.

Pandey, G.S. and Chunekar, K.C. (1999). *Bhavaprakasa Nighantu* (Indian Materia Medica) of Sri Bhavamisra (c. 1500-1600 AD). Chaukhamba Bharati Academy, Varanasi 221 001, India. 984 pp.

Peters S.E. (1994). Conversion to low –input farming systems in Pennsylvania, USA: An evaluation of the Rodal Framing Systems Trial and related economic studies. In: Lampkin NH and Padel S (eds). Economics of organic farming. CAB, Wallingford,UK. pp. 265-284.

Pillay, J.K. (1972). Educational system of the ancient Tamils, Madras.

Powell D. (2000). Update: potential impact of pollen of bt- CORN. http://www.foodsafety.ksu.edu/articles/705/bt_corn_impacts.pdf

Rajendran T.P., Venugopalan M.V. and Tarhalkar P.P. (2000). Organic cotton farming in India. Central Institute of Cotton Research. Technical Bulletin No. 1/2000, Nagpur, p. 39.

Ramesh P., Singh M. and Rao A.S. (2005). Organic farming: Its relevance to the Indian context. *Current Science India*, 88(4): 561-568.

Ray P.K., Prasad A.K. and Nandan, R. (1985). Pesticides – environmental problem. *Science and Culture*, 57: 363–371.

Raychaudhuri, S.P. 1964. *Agriculture in Ancient India*. Indian Council of Agricultural Research, New Delhi, India. 167 pp.

Rodda J.C. and Ubertini L (2004). The Basis of Civilisation-Water Science? International Association of Hydrological Science 2004, IAHS Press, Centre of Ecology and Hydrology, Wallingford, Oxfordshire, OX 10 8BB, UK.

Roundup Ready Soybeans, http://www.biotechknowledge.monsanto.com/biotech/knowcenter.nsf

Rundlof M., Edlund M. and Smith H.G. (2010).Organic farming at local and landscape scales benefits plant diversity. Ecography (In press); DOI: 10.1111/j.1600-0587.2009.05938.x.

Rundlof M. and Smith H.G. (2010). Effectiveness of an agri-environment scheme, organic farming, in uniform agricultural landscapes. Basic and Applied Ecology in review.

Sadhale, Nalini. (Tr.) (1996). Surapala's *Vrikshayurveda* (The Science of Plant Life) by Surapala. *Agri-History Bulletin No.1*. Asian Agri-History Foundation, Secunderabad 500 009, India. 104 pp.

Sadhale, Nalini. (Tr.) (1999). *Krishi-Parashara* (Agriculture by Parashara). *Agri-History Bulletin No.2*. Asian Agri-History Foundation, Secunderabad 500 009, India. 104 pp.

Shamasastry, R. (Ed.).(1926). *Abhilashitarthchintamani* of Someshwara Deva. Bhudharakrida (V - I). Government Branch Press, Mysore, India.

Shamasastry, R. (1961). *Kautilya's Arthasastra*. Seventh Edition. Mysore Printing and Publishing House, Mysore, India. 482 pp. (First Edition published in 1915.)

Singh V.P. and Yadava R.N. (2003). Water Resources System Operation: Proceedings of the International Conference on Water and Environment. Allied Publishers, New Delhi, India.

Sivarajan, V.V. and Balachandran, Indira. (1994). *Ayurvedic Drugs and their Plant Sources*. Oxford & IBH Publishing Co., New Delhi, India. 570 pp.

Stockdale E.A., Lampkin N.H., Hovi M., Keatinge R., Lennartsson E.K.M., Macdonald D.W., Padel S., Tattersall F.H., Wolfe M.S. and Watson C.A. (2000). Agronomic and environmental implications of organic farming systems. *Adv. Agron.*, 70: 261– 327.

Stolze M., Piorr A., Haring A. and Dabbert S. (2000). The environmental impact of organic farming in Europe. In Organic Farming in Europe: Economics and Policy. University of Hohenheim, Hohenheim,Germany.

Thulasamma, L. (2006). Technical Change in Agriculture, 1952–2000. Encyclopedia of India (vol. 4) edited by Stanley Wolpert, pp. 143–148, Thomson Gale, ISBN 0-684-31353-7.

Van Bruggen A.H.C. (1995). Plant disease severity in high-input compared to reduced-input and organic farming systems. The American Phytopathological Society, St. Paul, Minnesota.

Vanderschuren H. and Zhang P. (2011).The BioCassava Plus Program: Biofortification of Cassava for Sub-Saharan Africa. *Annual Review of Plant Biology*, 62: 251–272.

Whitman D.B. (2000). Genetically modified foods: harmful or helpful?. CSA Discovery guides.1- 13.

Woese K., Lange D., Boess C., Bogl K.W. (1997). A comparison of organically and conventionally grown foods – Results of a review of the relevant literature. Journal of the Science of Food and Agriculture, 74: 281–293.

Index

S

T

V

W

Y

Z

www.ingramcontent.com/pod-product-compliance
Ingram Content Group UK Ltd.
Pitfield, Milton Keynes, MK11 3LW, UK
UKHW021947270726
14060UKWH00002B/397